Broader Impacts of Science on Society

How do scientists impact society in the twenty-first century? Many scientists are increasingly interested in the impact that their research will have on the public. Scientists likewise must answer this question when applying for funding from government agencies, particularly as part of the "Broader Impacts" criterion of proposals to the US National Science Foundation (NSF). This book equips scientists in all disciplines to do just that, by providing an overview of the origins, history, rationale, examples, and case studies of broader impacts, predominantly drawn from the author's experiences over the past five decades. Beyond including theory and evidence, it serves as a "how to" guide of best practices for scientists. Although this book primarily uses examples from NSF, the themes and best practices are applicable to scientists and applications around the world where funding also requires impacts and activities that benefit society.

Bruce J. MacFadden is a Distinguished Professor at the University of Florida. He has written 200 peer-reviewed publications and a book (*Fossil Horses: Systematics, Paleobiology, and Evolution of the Family Equidae*, Cambridge University Press, 1992). He has been Principal Investigator of 50 NSF grants totaling more than $35 million over the past five decades. Professor MacFadden is a Fellow of the American Association for the Advancement of Science, the Geological Society of America, and the Paleontological Society. He was President of the Society of Vertebrate Paleontology (1986–1988) and President of the Paleontological Society (2018–2020). A former NSF Program Officer (2009–2010), he teaches graduate seminars and provides professional development on Broader Impacts.

"MacFadden has led an extensive life in science, as a paleontologist, museum curator, university administrator, and National Science Foundation staff member. His analyses of NSF history and policy changes from the agency's 1952 start through 2018 – and of many successes and challenges of his own – will be invaluable to anyone seeking research funds from this important government entity. The book focuses on explaining NSF's poorly understood Broader Impacts requirement, and it is especially needed now, when in some programs only about one in ten applications to NSF for funding are successful."

– Bruce Alberts, University of California, San Francisco;
former President of the National Academy of Sciences

"An excellent, pragmatic guide to the philosophy and practice of articulating the many dimensions of broader impacts of science on society, from a highly respected and experienced paleontologist and former National Science Foundation Program Officer. Well-written and accessible, this worthwhile book provides clear and useful information and advice on planning, preparing, and executing activities that are motivated by the genuine spirit of achieving a broader societal impact beyond scientific research. As it becomes increasingly important for scientists to communicate more often and more effectively with non-scientists, *Broader Impacts of Science on Society* serves as an invaluable resource to all who seek to extend the reach of their specific scientific insights and expertise to the rest of the world."

– Sandra J. Carlson, University of California, Davis;
Past-President of the Paleontological Society

"An accessible and practical book that models a reflective and informed approach to thinking about the impacts of scientific research on society. Those who design, study and evaluate Broader Impacts, communication, and learning activities will find the examples, anecdotes, and background information very relevant and useful. Having been on many sides of the audience equation, MacFadden provides a holistic guide for enlightened, engaged scholarship and practice of Broader Impacts."

– Jamie Bell, Project Director, Center for Advancement
of Informal Science Education (CAISE)

"This book provides an in-depth look at all aspects of broader impacts. It is a resource for anyone interested in the historical development of Broader Impacts as well as those seeking to understand the complexity of the Broader Impacts criterion and how to effectively address it."

– Susan Renoe, University of Missouri

Broader Impacts of Science on Society

BRUCE J. MACFADDEN
Florida Museum of Natural History
University of Florida, Gainesville

CAMBRIDGE
UNIVERSITY PRESS

University Printing House, Cambridge CB2 8BS, United Kingdom

One Liberty Plaza, 20th Floor, New York, NY 10006, USA

477 Williamstown Road, Port Melbourne, VIC 3207, Australia

314–321, 3rd Floor, Plot 3, Splendor Forum, Jasola District Centre,
New Delhi – 110025, India

79 Anson Road, #06–04/06, Singapore 079906

Cambridge University Press is part of the University of Cambridge.

It furthers the University's mission by disseminating knowledge in the pursuit of
education, learning, and research at the highest international levels of excellence.

www.cambridge.org
Information on this title: www.cambridge.org/9781108421720
DOI: 10.1017/9781108377577

First published 2019

Printed in the United Kingdom by TJ International Ltd. Padstow Cornwall

A catalogue record for this publication is available from the British Library.

Library of Congress Cataloging-in-Publication Data
Names: MacFadden, Bruce J., author.
Title: Broader impacts of science on society / Bruce J. MacFadden (Florida
Museum of Natural History, University of Florida, Gainsville).
Description: Cambridge ; New York, NY : Cambridge University Press, 2019. |
Includes bibliographical references and index.
Identifiers: LCCN 2019009159 | ISBN 9781108421720
Subjects: LCSH: Science – Social aspects. | Science – Study and teaching –
Social aspects.
Classification: LCC Q175.5 .M2827 | DDC 303.48/3–dc23
LC record available at https://lccn.loc.gov/2019009159

ISBN 978-1-108-42172-0 Hardback
ISBN 978-1-108-43428-7 Paperback

Contents

Preface

In the classic report *Science: The Endless Frontier*, Vannevar Bush (1945) envisioned the future of US science and technology for the second half of the twentieth century. This report also provided a roadmap for the creation of the National Science Foundation (NSF) in 1950. In a modern context, science, technology, engineering, and mathematics, commonly referred to as "STEM," are in many respects fundamental drivers of globalized nations as they advance in the twenty-first century. Despite the importance of STEM, however, professionals working in these fields have traditionally not done a good job of communicating their knowledge and discoveries outside the comfort zone of their content discipline. This lack of proper communication and dissemination has unfortunate outcomes with respect to both public understanding and support for basic research. This is where Broader Impacts come into focus. Although more context for the notion of Broader Impacts is provided throughout this book, the foundation of this concept is embodied within the desire for relevance to and benefit of science and STEM for society.

The purpose of this book is thus to provide a background, overview, and guide to the range of Broader Impacts and related activities, particularly as these relate to NSF. These themes are intended to serve as examples and best practices for professionals wanting to reach out to society. For reasons that will hopefully become clearer in what follows, two terms are used throughout this book. "**Broader Impacts**" is used in the context of the merit review criteria and activities that one does as part of a project funded by NSF. In contrast, when I use "**societal benefit**," "**benefit society**," or "**benefit to society**," these represent a more inclusive context that encompasses activities that may, or may not, include specific reference to NSF (Inset 0.1). For example, one might develop a personal website that describes their research and outreach activities. In this context, this is an example of outreach for **societal benefit**, with a goal of disseminating knowledge and activities to a large audience. Perhaps this same scientist then submits an NSF proposal for a new research idea; in this case they could repurpose their website as part of a **Broader Impacts** plan within the proposal description.

Science versus STEM

When NSF was founded in 1950, the acronym STEM did not exist; this word has, however, become commonplace in society today (Inset 0.1). As a scientist, I consider myself to be part of STEM. In the chapters of this book, I sometimes use *science* and *STEM* interchangeably. This is similar to the broad use of the term science (e.g., Bush's *Science: The Endless Frontier* [1945], which was meant to also include technology, and if he were writing today, likely STEM). Likewise, the name "National Science Foundation" includes the word science, but in the present-day context this institution encompasses STEM as well. I considered whether "STEM" or "science" should be in the title of this book, and went back and forth several times.

> ### Important words and phrases used in this book
>
> - **STEM**: Science, Technology, Engineering, and Math; term used over the past few decades to include the four words that form the acronym.
> - **Science**: Can be used in a more restrictive context to mean only the science in STEM, or in a more inclusive context to include all STEM activities. For example, the activities of the National Science Foundation (NSF) encompass STEM. The words "science" and "scientist" are frequently used in a more inclusive context in this book.
> - **Broader Impacts**: Specific activities proposed for NSF projects.
> - **Societal benefit**: the goal of scientists who do outreach activities, some of which are supported by NSF, others of which may be done independently.

Inset 0.1

Some of my colleagues, particularly in education, find STEM to be a hackneyed term that may at some point outlive its identity (e.g., STEAM [Art] is trending now). In addition, as the book developed, given my background as a scientist, after Chapter 1 most of the examples that I use come from science. If I had used STEM in the book title it might have misled the reader. In the final analysis, whereas STEM is commonplace now, "science" is a more enduring term that will likely persist, and the latter better identifies the primary focus of this book. Thus, in the pages below, the terms "science" and "scientist" are frequently used to also include the other aspects of STEM.

Audience

When I worked as a program officer at NSF (2009–2010), a colleague taught me that one of the most important things that needs to be understood is to know your intended audience (Chapter 8). While this seems obvious, it is not always understood by scientists. I try to know my audience, not just when writing proposals, but also during classes, academic seminars, publishing peer-reviewed papers, or presenting public talks. Thus, we should be aware of our audiences any time we communicate with them. Audiences can be complex and oftentimes include intersecting subsets of people with different backgrounds and experiences. They are not necessarily homogeneous and invariably include learners with different knowledge. This is the case for this book as well.

I therefore see multiple intersecting target audiences for this book. The two primary ones include scientists wanting to: (1) do **Broader Impacts** – that is, to write a potentially more competitive NSF proposal; and/or (2) **benefit society** through dissemination, education, and outreach programs. Most of the audience, and indeed my emphasis, relates to audiences in the United States. I also, however, have periodically been asked to review grant proposals in other countries. Although

Broader Impacts may be called by some other term in other countries, in many instances equivalent expectations exist for professionals to provide plans and activities for general societal benefit. Thus, although primarily focused on the United States, this book might also be useful to scientists in other countries. I also hope that this book might be of interest to people outside of academe who would like to understand ways in which they can connect with scientists and participate in mutually beneficial activities and programs.

Intent, Scope, and Structure

As the name implies, the topic of this book – *Broader Impacts of Science on Society* – is potentially expansive. When I think of the scope of the table of contents, it is both daunting and unrealistic, particularly for one person to try to tackle. Thus, the emphasis of this book is Broader Impacts and societal benefit based on my experiences. While the goal for this book is to be comprehensive, that is, covering most of the relevant topics, its scope is not intended to be exhaustive. Many of the chapters could easily be books in themselves. In fact, some of them, like communications – on which I am neither an expert nor an authority – are entire disciplines for which the chapter is only scratching the surface. Nevertheless, these topics are important to cover within the overall theme of this book and are therefore included here. There are inherent emphases in how I treat the various topics. As a paleontologist, evolutionary biologist, and museum scientist, I tend to focus on the natural sciences because these are closer to my knowledge base and content domain. Nevertheless, science and STEM in the twenty-first century seek to be integrative and this book is mindful of this trend.

Because this book is based on my experiences, I oftentimes write in the first person. This is not always best practice, particularly in formal scientific writing. Nevertheless, this book is intended to be my view of the subject. I also believe that stories are engaging narrative. As such, each chapter starts with a first-person anecdote. In other sections the more widely accepted third person is mostly, but not always, used, particularly for descriptive narrative and discussion of relevant background and literature.

Most of the chapters have a similar structure. They typically include: (1) a personal anecdote; (2) some background, such as from previous studies or learning research; (3) at least one representative example or case study; and (4) a discussion of relevant best practices and broader significance. Likewise, although there is a lengthy list of about 600 references cited, in most cases these are just scratching the surface.

This overall intention of this book is to help scientists understand and develop Broader Impacts activities and programs for the benefit of society. The focus is based on my experiences over the past five decades. While not intended to be a textbook, its structure, and many of the ideas and examples presented here, resulted from a graduate seminar on Broader Impacts that I have taught at the University of Florida since 2006. The number and sequence of chapters could realistically be covered in a four-month-long academic semester, either in an organized course or perhaps during the ubiquitous research "lab" meetings.

My Background and Acknowledgments

I have benefitted greatly because of others and formative experiences that I have had. As a child, I developed an abiding interest in dinosaurs. To me there was no more interesting place than the dinosaur halls at the American Museum of Natural History in New York, located just a train ride from where we lived in the suburbs. I truly believe in the profound impact that inspirational K–12 teachers can have on young minds. My tenth-grade earth science teacher, Mr. Greenstein, at Port Chester Senior High School (New York), encouraged my interest and built my confidence in science (Chapter 10). For his class I wrote a term paper on the fossil bird *Archaeopteryx*. I enjoyed going to the Yonkers Public Library on Saturdays, where I did the research for this paper.

As an undergraduate at Cornell University, I had the opportunity to take school groups on field trips along the shores of Lake Cayuga to collect "elephant toenails," which were fascinating 375 million-year-old fossil corals. These outings with young learners turned out to be a particularly rewarding experience. One of the joys of teaching is to see students enjoying learning and thrilled by discovery. At that time, I was also mentored by two influential Cornell geology professors, John Wells and Art Bloom, who continued to build my confidence and encouraged me to consider a career as a professional scientist.

As a graduate student at Columbia University in the early 1970s I did what most graduate students were supposed to do, or at least I think I did. My major professor, Malcolm McKenna, taught me to think and work independently. He always encouraged us to do the best science that we could and to think big thoughts – "outside the box." There was no such thing as NSF's Broader Impacts then, so we were primarily focused on doing good science and getting a job upon graduation. With regard to the latter, in 1976 I was fortunate to land an instructorship in the geology department at Yale University, which was adjacent to the Peabody Museum of Natural History (also part of Yale). During that time the museum was renovating some of its exhibit halls. Given my prior research interests as a graduate student, I was invited to participate in an update of the fossil horse display. With the help of the museum design and fabrication staff, I became involved in the update. I enjoyed the creative process of designing the exhibit and preparing its interpretive materials. In hindsight, this was the beginning of my professional interest in informal science education, now also called informal STEM learning.

Since 1977 I have been a professor and faculty curator at the Florida Museum of Natural History, which is part of the University of Florida. This position has come with the expectation that we participate in the education and outreach activities of our museum. It has been both immensely rewarding and creative to work on exhibits and public programs. In the 1990s, when we were building a new education and exhibits center, I was given the opportunity to take a break from my full-time faculty duties to be involved in the administration of exhibits and public programs. From

Fig. 0.1 *Hall of Florida Fossils* at the Florida Museum of Natural History, University of Florida (Mary Warrick photo).

1996 to 2004 I was in charge of the development of this museum facility and its programs. Despite its administrative challenges, this was highly creative and rewarding. It also provided the opportunity to communicate science through exhibits, such as in the Hall of Florida Fossils, a space that still has special meaning to me (Fig. 0.1). During this time I started reviewing NSF proposals in informal science education. These were a refreshing and innovative change from the typical science research proposals that I had reviewed for decades. It also was the same time that Broader Impacts were starting to ramp up as part of the merit review criteria for NSF proposals.

I was involved in the development of this exhibition during my time as director of the Exhibits and Public Programs division. The skeleton in the foreground of Fig. 0.1 represents the Ice Age (Pleistocene) horse *Equus* surrounded by skeletons of other evidence of past life in Florida. *Equus* has special significance because of my long-standing research interest in fossil horses (e.g., MacFadden, 1992). It also exemplifies another interest of mine – the translation of research into public exhibits and how fossil horses provide evidence for evolution (MacFadden et al., 2012).

In the mid-2000s, I become more aware of NSF's Broader Impacts and started to teach a graduate seminar on this subject (MacFadden, 2009). I have taught this course for seven semesters since that time. I thank my students in these classes for their enthusiasm. Unlike some of my more senior colleagues, who are set in their ways, the next generation "gets" NSF's Broader Impacts and related activities that benefit society. I have periodically co-taught this course with my colleague David Reed, and thank him for his insight, perspective, and shared enthusiasm.

In 2009, after participating in an NSF panel, I was encouraged to apply for, and accepted, a position as a "rotating" (fixed-term) program officer at NSF in the Directorate of Education and Human Resources (EHR). This opportunity placed me in the informal science education program the Lifelong Learning Cluster. During that year (2009–2010) at NSF I managed proposals and projects mostly related to science learning in the built environment (i.e., museums, science centers, planetaria, aquaria, nature centers, and the like). At that time NSF had a collaboration with the American Association for the Advancement of Science (AAAS) to promote science communication via workshops at meetings and universities. I represented NSF at these events by presenting talks on Broader Impacts. I am grateful for how that year at NSF broadened my professional development. I also gained an insider's view on how committed program officers are to NSF's mission and the hard work they do. As professors, we like to think that we are dedicated and work hard; I saw the same culture at NSF. I greatly enjoyed my time there; it was critical to my thinking about Broader Impacts. There are so many colleagues there to thank, including Al DeSena, my de facto mentor, fellow rotators Sue Allen and Leslie Goodyear, and Wyn Jennings, the last of whom has encouraged me since I left NSF a decade ago.

As described in Chapter 11, my recent passion has been working with K–12 STEM teachers. The catalyst for this new direction in my career is an example of serendipity in science. After returning from NSF in 2011, I was contacted by Gary Bloom, then superintendent of Santa Cruz City Schools (California). Gary has a keen interest in fossils and paleontology. He had an enjoyable experience volunteering in the paleontology laboratory at the Smithsonian Tropical Research Institute (STRI) with Carlos Jaramillo. Gary wanted his teachers to have a similar experience. At the same time I was involved in a multi-year NSF-funded project to advance understanding of the geology and paleontology along the Panama Canal during its expansion. Carlos put Gary in contact with me. Together we wrote a series of successful NSF grants for STEM teachers to engage in the collection of fossils in Panama. We did this for five years and involved 50 teachers and three dozen scientists. I found that I enjoyed working with teachers and, as a result of this, in 2015 I applied for an "alternative sabbatical" from the University of Florida. These sabbaticals were intended for proposals that were "outside of the box," something different from the traditional hiatus, where a scientist spends time in a colleague's laboratory away from their home institution. During my alternative sabbatical in 2015–2016, I was a visiting scientist in the Santa Cruz County Office of Education (SCCOE). I am grateful to the superintendent, Michael Watkins, for allowing me to do my sabbatical at SCCOE. I am also grateful to Mary Anne James and Adam Wade for mentoring me in the K–12 ecosystem, and for so many teachers who welcomed me into their classrooms. Being a visiting scientist was a particularly rewarding part of my continued professional development. I found that I really enjoy teaching third-graders about local fossils. This has led to my newfound commitment to impacting society through K–12 outreach, education, and partnerships.

I am also grateful to my graduate students, who over the past decade have shown me that the next generation has no problem with Broader Impacts and related activities of societal benefit. To them it simply makes sense, and they are predisposed to the importance of these activities with little, if any, prodding. Students are an inspiration to me and should also be to some colleagues of my generation who still disparage these activities as taking away from their research. Much of what I have experienced over the past 40 years could not have been possible without the continued support of my home institution, the Florida Museum of Natural History at the University of Florida. I am likewise grateful for the support that I have received from grants, primarily from NSF.

While writing this book I benefitted greatly from discussions I had with colleagues and friends, most notably Douglas Jones and David Evans, both of the University of Florida. I also thank Jeff Gage and Leigh Anne McConnaughey of our photography department, who helped to organize and find photos. With much patience, Tammy Fluech cheerfully prepared most of the graphic illustrations. Many kind people, too numerous to mention here, responded to requests for images, permissions, and other information. My graduate students have remained interested and supportive, particularly during our lab meeting conversations. I thank Matt Lloyd, editor at Cambridge University Press for his insight, guidance, and patience. Zoë Pruce, also at Cambridge University Press, provided much helpful input during preparation and production of the manuscript. David Evans peer-reviewed the entire manuscript. David Jennings and Shari Ellis critically read, respectively, Chapters 17 (Project Management) and 18 (Evaluation). My student Jeanette Pirlo carefully edited early drafts of all the chapters and then did the final check of references. She showed me that even though I have written lots during my career, I have a different interpretation of the use of commas and oftentimes write in compound, run-on sentences.

The most important influence in this journey has been my lovely wife, Jeannette, who has encouraged me over the past quarter-century. She has validated that my interest in Broader Impacts and related activities is a worthy way for a scientist to contribute to society. Jeannette has been entirely supportive of me while I wrote this book, which has consumed many early mornings and evenings at home and during "workations."

In closing, I want to acknowledge the many other persons who have influenced me. I hope that although they are not mentioned by name above, they will accept my gratitude as having been influential in my professional development. I would not be where I am today professionally without them.

Abbreviations, Definitions, and Acronyms

AAAS	American Association for the Advancement of Science
ABR	accomplishment-based renewals, NSF
ADA	Americans with Disabilities Act
ADBC	Advancing Digitization of Biodiversity Collections, NSF
AISL	Advancing Informal STEM Learning, NSF
AMNH	American Museum of Natural History
AP	advanced placement
AR	augmented reality
ARRA	American Recovery and Reinvestment Act
ASTC	Association of Science and Technology Centers
ATE	Advanced Technological Education, NSF
BIO	Directorate for Biological Sciences, NSF
BOP	Billion Oyster Project
BP	Broadening Participation, NSF
CAISE	Center for Advancement of Informal Science Education
CCEP	Climate Change Education Partnership, NSF
CoP	community of practice
CRPA	Communicating Research to Public Audiences, NSF
DCL	Dear Colleague Letter, NSF
DEI	diversity, equity, and inclusion
DGE	Division of Graduate Education, NSF
DORA	Declaration on Research Assessment
DRL	Division of Research on Learning, NSF
DUE	Division of Undergraduate Education, NSF
EHR	Directorate for Education and Human Resources, NSF
ELL	English language learners
EPSCoR	Experimental Program to Stimulate Competitive Research, NSF
ESL	English as a second language
FCAT	Florida Comprehensive Assessment Test
FHC	Fossil Horses in Cyberspace
FLMNH	Florida Museum of Natural History, UF
GABI	Great American Biotic Interchange
GBIF	Global Biodiversity Information Facility
GEO	Directorate for Geosciences, NSF
GRFP	Graduate Research Fellowship Program (also GRF), NSF
HRD	Division of Human Resource Development, NSF
iDigBio	Integrated Digitized Biocollections

IGY	International Geophysical Year
IMLS	Institute of Museum and Library Services
INCLUDES	Inclusion across the Nation of Communities of Learners of Underrepresented Discoverers in Engineering and Science, NSF
IRB	Institutional Review Board
ISE	Informal Science Education
ISL	Informal STEM learning
ITEST	Innovative Technology Experiences for Students and Teachers, NSF
JIF	journal impact factor
K–12	Schools in the United States (Kindergarten through grade 12), and encompassing elementary, middle, and high school
K–16	K–12 plus four years of undergraduate college or university education
MSP	Math and Science Partnerships, NSF
MACOS	"Man: A Course of Study"
NAS	National Academy of Sciences; also National Academies of Sciences, Engineering, and Medicine
NGSS	Next Generation Science Standards
NIH	National Institutes of Health
NOS	nature of science
NPS	National Park Service
NRT	NSF Research Traineeship
NSB	National Science Board
NSDL	National Science Digital Library
NSF	National Science Foundation, United States
NSTA	National Science Teaching Association
OA	open access
OIA	Office of Integrative Activities, NSF
OISE	Office of International Science and Engineering, NSF
PD	professional development, mostly in relation to K–12 teachers
PI	principal investigator
PIRE	Partnerships for International Research and Education, NSF
PPSR	public participation in scientific research
R&D	research and development
R&RA	Research and Related Activities, NSF
RANN	Research Applied to National Needs
RCN	Research Coordination Network, NSF
RET	Research Experiences for Teachers, NSF
REU	Research Experiences for Undergraduates, NSF
ROI	return on investment
SBE	Directorate for Social, Behavioral and Economic Sciences, NSF
SBIR	Small Business Innovation Research, NSF
SCCOE	Santa Cruz County Office of Education, California
SEES	Science, Engineering, and Education for Sustainability, NSF

SGER	Small Grants for Exploratory Research, NSF
SMP	Science Masters Program, NSF
STC	Science and Technology Centers, NSF
STEAM	STEM plus art
STEM	science, technology, engineering, and mathematics
STRI	Smithsonian Tropical Research Institute, Panama
TACC	Texas Advanced Computing Center, University of Texas
TCN	Thematic Collections Network, NSF
TED	Technology, Entertainment, Design
UF	University of Florida
VR	virtual reality
VSA	Visitor Studies Association

1 Introduction: Science, STEM, and Society

Why Do You Want to be a Scientist?

I have taught a graduate seminar on Broader Impacts since 2006. On the first day, I open the class with the simple question above. It appears from the quizzical looks on many students' faces that they have never been asked this question, or even considered it for themselves. This notable lack of self-reflection and awareness is despite the fact that this fundamental question defines who they are and what they are studying to be. As a student, I likewise never considered this question or its significance.

Now that I have asked this question of many students, I can predict two kinds of responses. In the first, the student describes their innate curiosity and how they want to understand the natural world. In the second, the student describes how they want to make the world a better place, or to have a positive impact on society. This latter response is consistent with the notion of social responsibility (Inset 1.1). It also becomes clear after several responses that some of the students are interested in both – they have innate curiosity, as well as a desire to find meaning in what they will do with their lives.

I then ask the students why they are taking this course. They typically respond that they are interested in the topics, or some say that they are taking it to make them more successful with grants, particularly from NSF. Some students are taking the class despite the skepticism of their major professors. Many of their mentors were brought up in an ivory-tower culture devoid of Broader Impacts. As such, they typically lack an appreciation of the importance of societal benefit in their research. Thankfully, the culture is changing.

On an upbeat note, as exemplified by the students in my class, most of the next generation are far more accepting of the philosophical

---

Social responsibility

Ethical framework in which individuals and organizations are expected to act for the overall benefit of society.

NSF's Broader Impacts exemplify the social responsibility of science.

Inset 1.1

justification for why they are training to be scientists. NSF's Broader Impacts are therefore an easier sell to the next generation, and this bodes well for the future. Social responsibility and societal benefit are not concepts particular to the United States, but they are part of the ethical framework in other countries that support basic research (e.g., Rajput, 2018).

Introduction

New frontiers of the mind are before us, and if they are pioneered with the same vision, boldness, and drive with which we have waged this war we can create a fuller and more fruitful employment and a fuller and more fruitful life.

Franklin D. Roosevelt, 17 November 1944 (in Bush, 1945)

This chapter provides an overview of why science, and in a broader context STEM, are of fundamental importance to the progress of nations and their citizens in the twenty-first century. We will return to some of the topics in subsequent chapters, but here they provide the foundation and rationale for the fundamental importance of science and STEM in society. The focus of this chapter is the context of science and STEM in the United States, primarily during the second half of the twentieth century and beginning of the twenty-first century. Nevertheless, in a globally connected world, much of what is described here also pertains to other STEM-enabled countries as well.

Most of this chapter is guided by a 1945 report titled *Science: The Endless Frontier*. It was transmitted to US President Franklin D. Roosevelt by Vannevar Bush (1945), who at the time was Director of the Office of Scientific Research and Development. Much of the context of that report relates to the enormous impact that science had on the development of technology and practical applications during World War II. Several quotes from this report presented in the following exemplify the vision and framework for the next half-century. Although the acronym STEM had not yet been invented, this report charted the course for the development of science, technology, engineering, and mathematics in the United States after World War II. In the early part of the twenty-first century, Bush's vision and recommendations still have relevance (Pielke, 2010).

This chapter concludes with a discussion of politics and science in the twenty-first century. Bush understood that for science to be successful, the federal government would need to invest in this enormous enterprise. A corollary is that oversight and accountability would be important. It is unclear whether Bush, or any other learned person in the middle of the twentieth century, could have foreseen the extent to which science and STEM has been "politicized" in modern society. This is exemplified by politicians taking stands on "hot-button" topics such as evolution, climate change (*sensu*

Leshner, 2010), and federal versus state standards dictating how STEM should be taught in K–12 schools.

Value of Science and STEM in Society

The rationale for science that Vannevar Bush (Inset 1.2; Carnegie Institution of Washington photo) envisioned in the middle of the twentieth century still holds true in the context of how STEM is of value in service to modern society. His words in 1945 were heavily influenced by the context of military superiority during World War II. This is not surprising, given the fact that many important innovations, emerging fields, and new ways of doing science and technology occurred during this time, funded by investments from government and the private sector. To name just a few: radar (and sonar), computers, and airplanes all saw major leaps forward during World War II and thereafter were put to use for the benefit of society. The field of operations research largely started with British and

"...without scientific progress no amount of achievement in other directions can insure our health, prosperity, and security as a nation in the modern world."
– Vannevar Bush

Inset 1.2

American scientists during this time. For example, Patrick Blackett (who later received the Nobel Prize) used applied mathematics to optimize the configuration of anti-aircraft installations defending London, or the size of merchant ship convoys crossing the North Atlantic to defend against German U-boat attacks (Budlansky, 2013). The Atomic Age, of huge societal impact – both positive and negative – also started during this time. Many of the scientists of the Atomic Age also understood the importance of public understanding of science and technology for societal benefit – for example, in 1969 the physicist Frank Oppenheimer (brother of Robert J. Oppenheimer) founded the Exploratorium science center in San Francisco (Exploratorium, 2018).

It is also understood that science is important during peacetime. In addition to the advances in knowledge, innovation, and scientific capital, in modern society STEM employs 5–10 percent of the workforce in the United States (US Department of Commerce, 2011). Although not a large percentage of the overall workforce, these jobs (Fig. 1.1) are important for economic progress and world leadership. Many of these jobs require considerable levels of education and, in turn, reward workers in careers with appropriate levels of compensation. Moreover, the rate of job growth in the United States is greater in STEM than in non-STEM fields (Fig. 1.2).

Science is a catalyst for health information technology, agriculture, and energy, driving the national economy; it deserves to be valued as such.

Barbara Schaal, AAAS President (2017)

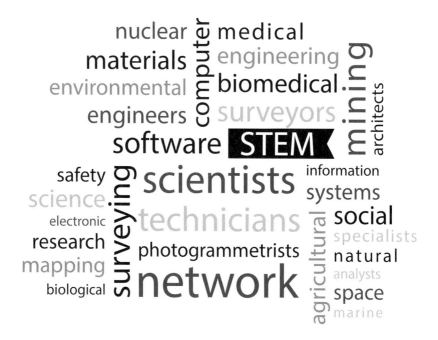

Fig. 1.1 Word cloud of STEM job code names (compiled from Noonan, 2017).

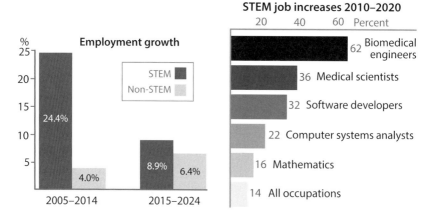

Fig. 1.2 Left: Projected employment growth in STEM versus non-STEM fields in the early twenty-first century (Noonan, 2017). Right: Projected increases in STEM jobs, by category, 2010–2020 (US Department of Education, 2017).

Since the middle of the twentieth century, a sense of importance has emerged among the world's industrialized nations about their dominance, or position, with regard to science and technology. Bush (1945) advocated that the United States should lead in this regard, rather than play catch-up from innovations produced by other nations. In a modern context, China is undergoing reform of its government's science and technology infrastructure with a vision to be a world leader rather than being in a position of technological catch-up. Government, industry, and academic leaders have come together to formulate a national research and development (R&D) agenda with a goal of improving China's position in science and technology through the middle of the twenty-first century (Cao & Suttmeier, 2017). China's recent investment in R&D is impressive and on the rise. China is ranked second behind the United States in overall R&D investment, but significant gains are being seen from the former. Whereas in 2012 China spent 34 percent as much as the United States on R&D, in 2017 it was closer to 45 percent. Thus, in China funding for this sector of the economy has risen significantly, by 12.3 percent (to the equivalent of $254 billion) in 2017 over the previous year, and it is likely to continue to rise (Normile, 2018).

In order to be a world leader in science and technology, nations need to invest in creative work that increases knowledge. One metric of this investment is the research expenditure of the government and private sectors as a percentage of the gross domestic product (GDP). Using this metric, at 2.7 percent of its GDP in 2013, the United States ranks ninth (China is fourteenth) of all countries for which data are available (Fig. 1.3), after Korea, Israel, Japan, Sweden, Finland, Denmark, Austria, and Germany (World Bank, 2017).

The world's population is projected to exceed nine billion by 2050 (Jarvis, 2017). At its current rate of growth, the US population is estimated to be 436 million by 2050 (Passel & Cohn, 2008); which represents a 34 percent increase from 325 million in 2017 (US Census Bureau, 2017a). This inexorable growth of the human species on

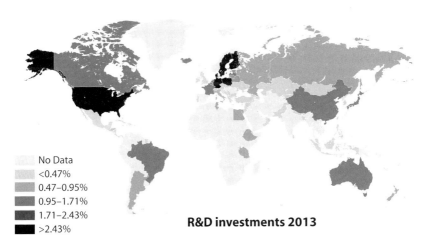

Fig. 1.3 Expenditures for R&D as a percentage of the GDP for 2013, by country (World Bank, 2017).

Earth further underscores the importance of STEM to address fundamentally important issues, including national security, health, food, water, energy, environment,
and sustainability in the twenty-first century.

Basic Science and "Useless Knowledge"

> Basic research leads to new knowledge. It provides scientific capital. It creates the
> fund from which the practical applications of knowledge must be drawn.
>
> Bush (1945)

It is easy to understand why scientific and technological innovation are important
when there is direct, or perceived, benefit for society. But how can we rationalize
investment in the production of knowledge for its own sake without immediate
application? This question has existed ever since governments have invested in
science. It is of particular importance in organizations, whether they be in government or the private sector, that are not mission-based, that is ones that support what
might be called basic research (Bush, 1945; Pielke, 2010). "Basic research" is done
for science's sake because it is potentially novel, innovative, and leads to new
discoveries. The same is also true for much of the research in mathematics. Unlike
basic science and mathematics, within the context of STEM, technology and engineering, as well as medical research, are typically identified with a problem to be
solved or a direct application in mind.

Three-quarters of a century ago, Bush understood and supported the idea of
basic research as an investment in the infrastructure and knowledge base of
the United States. He is not the only high-profile scientist of the middle
twentieth century who both understood and advocated for the place of basic
science in society. In 1939, Abraham Flexner, who founded the Institute of
Advanced Study at Princeton University (and was instrumental in bringing
Albert Einstein to the United States), wrote an essay entitled "The usefulness
of useless knowledge." In a recent retrospective about Flexner and his essay,
Dijkgraaf (2017; see also Tovey, 2017) describes not just how basic science is
available for future applications, but also its importance in social and cultural
change. As an example, without quantum mechanics there would be no
computers, fiber-optic telecommunications, or smartphones (Orzel, 2015).
Lasers were initially developed a half-century ago primarily within a framework of basic research (Cartlidge, 2018), but have subsequently been translated into many important applications that have benefitted society and
impacted the modern world.

In Chapter 2 we will learn in more detail about the beginning of the US National
Science Foundation (NSF) in 1950, which largely resulted from the advocacy of
Vannevar Bush. He and NSF's founders understood that science needed to be done
for a variety of reasons that would impact both basic knowledge as well as practical
applications. It is in the realm of basic science that the seeds of NSF's Broader

Impacts were sowed. Almost since the beginning of NSF, and particularly in a time of increased skepticism about science and STEM in society, there has been tension between "ivory tower" science and government oversight and accountability. The extent to which society values basic research is at the core of the debate about return on investment of government funds.

The debate about the value of pure science hits close to home because my field is paleontology, the study of fossils and what they tell us about evolution and the Earth during the past 4.6 billion years of "Deep Time" (McPhee, 1981). There are some notable exceptions, such as the new field of conservation paleobiology (Dietl & Flessa, 2011; 2018), in which studies of past ecosystem changes can inform modern policies about conservation. Otherwise, paleontology does not have much direct, or immediate, practical application. Nevertheless, the study of fossils informs relevant topics such as evolution and climate change. In addition, as many kids will tell you, dinosaurs are cool, and these extinct creatures therefore provide a charismatic gateway for STEM engagement in the twenty-first century. The direct societal benefit, if not practical application, of paleontology is that it potentially engages the next generation of STEM learners and hopefully opens their minds to discovery.

Diversity, Workforce, and STEM

> Studies clearly show that there are talented individuals in every part of the population, but with few exceptions, those without the means of buying higher education go without it. If ability, and not the circumstance of family fortune, determines who shall receive higher education in science, then we shall be assured of constantly improving quality at every level of scientific activity.
>
> Bush (1945)

Between 5 and 10 percent of the workforce in the United States, or about 8–9 million people in 2017, are employed in STEM jobs. It is clear that the key to successfully developing this workforce is to include the diversity represented in the United States, including men, women, and underrepresented minorities. A goal for this strategy should be that the STEM workforce mirrors the diversity of the United States as a whole. While this is a laudable and appropriate goal with regard to social responsibility, it is not the reality today. The STEM workforce in the United States, particularly in senior positions and higher-paying jobs, is disproportionately represented by white males.

The reasons underlying the lack of diversity within the STEM workforce in the United States are complex. This disparity is currently not being effectively addressed in order to achieve full equality. The two most affected groups in this regard include women and underrepresented minorities. Although women oftentimes are interested in STEM at an early age, for complex reasons they progressively lose interest both in STEM and in related careers. For those females who persevere past adolescence, the road to a career in STEM is no less easy (Handelsman et al., 2005; McNutt,

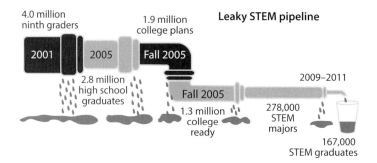

Fig. 1.4 The STEM leaky pipeline showing loss of women, minorities, and low-income students during different phases of the educational and workforce system (modified from Dubois, 2014).

2013). This attrition has been termed the "leaky pipeline" (Fig. 1.4), or the progressive loss of women and underrepresented minorities "up the ladder" and in more senior positions in STEM (Blickenstaff, 2005; Dubois, 2014). This loss is not just restricted to the United States, but also exists in other countries (Professionals Australia, 2017; Resmini, 2016). As an example of the disparity in the workforce, women researchers in STEM vary from about 40 percent in China to a low of 20 percent in other parts of Asia and the eastern Pacific; the world average is 30 percent. In the United States women comprise half of the national workforce, but only one-quarter are employed in STEM fields (Landivar, 2013; Wu, 2016).

In addition to the disparity with women in STEM, minorities including Hispanics and Latinxs, African-Americans, and Native Americans are likewise poorly represented relative to their percentages in the US population. Although these groups collectively represent nearly 30 percent of the US population, they comprise only 9 percent of the STEM workforce (Ferrini-Mundy, 2013). This situation has not improved significantly since the turn of the twenty-first century despite multifaceted efforts toward equality and inclusion (Hrabowski, 2011). Interest in STEM among the next generation of African-American and Hispanic youth has actually declined since 2001. Without effective interventions, the outlook for future engagement in the STEM workforce is likewise not an optimistic scenario (Bidwell, 2015). The reasons for poor participation of these underrepresented minorities in STEM are complex, but relate to a variety of factors including family and social norms, perceived value, lack of relevance, educational achievement, and a dearth of suitable role models in STEM in the United States. In contrast to this lack of increased participation, non-US students, particularly from China and India, represent most of the growth in earned doctorates in STEM in the United States in the twenty-first century. Many of these graduates, however, return to their home countries and therefore do not contribute to the STEM workforce in the United States (Hrabowski, 2011). Another related problem with the current STEM workforce is the disparity between supply and demand. Thus, in some sectors such as biology, the supply of PhDs greatly outnumbers the available jobs; in fact, there is an order of magnitude difference

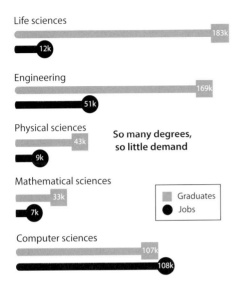

Fig. 1.5 Disparity between the supply of PhDs and demand for jobs within different STEM sectors (modified from Lohr, 2017).

between PhDs produced and available jobs (Lohr, 2017; Fig. 1.5). This hypercompetitive situation is exacerbated if the job search is focused on tenure-track positions in academia. Within the biomedical sciences, only 10 percent of trainees in the United States obtain the coveted tenure-track job five years after completing the PhD (Blank et al., 2017). In contrast, however, in computer science the supply of new PhDs roughly equals the supply of jobs. In a recent book, *The Graduate School Mess*, Cassuto (2016) analyzes this problem from the point of view of the challenging job market for PhDs in the humanities, but many of his observations and conclusions also pertain to certain sectors of STEM.

As Bush (1945) advocated, the diversity of minds, ideas, and human perspectives are fundamental to the success of the United States as a leader in scientific and technological innovation and as a player in the global market. From a pragmatic point of view, the majority of our future economic growth is linked to STEM. Until a workforce exists that more closely mirrors the demographics of the US population, and is optimized for supply and demand, then we will not have fulfilled our social responsibility or realize the economic benefits of diverse participation in STEM.

On to the Twenty-First Century

Globally Connected STEM

The twentieth century witnessed an unrivaled expansion of international conflicts, particularly during the two World Wars. During peacetime, international cooperation, diplomacy, and scientific collaboration were unprecedented. During the late twentieth century and the beginning of the twenty-first century, a globally connected STEM community has greatly benefitted from the technological innovations provided by cyberinfrastructure, most notably including electronic communication (email), social

Global competence

- International awareness
- Appreciation of cultural diversity
- Proficiency in foreign languages
- Competitive skills

Twenty-first-century skills
- Learning and innovation
- Digital literacy
- Career and life

Inset 1.3

media, and other cyberenabled modes of communication such as videoconferencing. These connections have done much to break down barriers of geography, time, and politics. Thus, STEM today involves international collaboration at an unprecedented rate. As an example, the number of peer-reviewed articles produced globally with authors from different countries has doubled since 1990 (Bollyky & Bollyky, 2012). This globally connected community also requires, and in turn benefits from, investments in political, economic, and social development made by both governments and the private sector.

Globally connected STEM also requires an expanded set of competencies and skills in order to optimize success in the twenty-first century (NEA, 2017; P21, 2017; Inset 1.3). The application of these competencies is also relevant to other segments of society, including business and government. Of relevance here, the competencies focus on international and cultural awareness and diversity (NEA, 2017). Despite the notion that English is the de facto language of science (van Weijen, 2012), global competencies should transcend the research and development infrastructure of STEM to include other languages as well. Only about 7 percent of the world's population of 7.5 billion people speak English as their native language. If percentages were a primary factor, then the language of science should be either Chinese (19 percent, 1.4 billion people) or Hindi-Urdu (8 percent, 588 million people), although English is spoken in more than 100 countries (Chinese is spoken in 33 countries; Noack & Gamio, 2015).

A generation ago, PhD students in the United States were required to demonstrate proficiency in one, or even two, foreign languages as part of their degree requirements. (I had French and German, but should have taken just Spanish instead.) Despite the other requirements placed on demanding PhD programs, it is beneficial for students actively involved in international research to communicate in the language of the collaborating country. In the twenty-first century, foreign-language competency for PhD students in US universities is not mandatory, although perhaps this requirement should be reinstated.

I was once invited to become a member of a PhD committee for a student conducting research in South America. He would be doing field work in rural areas, and Spanish would have been very helpful, and also shown the local scientists that he cared enough to learn the native language. I insisted that if I were to serve on his committee, the student would have to demonstrate proficiency in Spanish. His major professor said it was not necessary. I disagreed and the outcome was that I did not serve on this committee, which was fine with me.

In addition to global competence, leaders from diverse segments of society have realized the importance of what are now called twenty-first-century skills (P21, 2017; Trilling & Fadel, 2012; Inset 1.3). While much literature has been devoted to these skills within the context of advancing K–12 education, they are also widely used in business, community, and government. Traditionally, content expertise (i.e., knowledge and recall about a disciplinary domain such as paleontology) has been emphasized in education. Many of the twenty-first-century skills require higher-order cognitive processes, and harken back to classic studies like Bloom's taxonomy (Bloom et al., 1956). The twenty-first-century skills, however, place further emphasis on "soft skills," which are personality traits, goals, motivations, and preferences valued in the labor market, school, and many other domains. Although difficult to measure, the importance of soft skills relative to disciplinary "hard skills" is significant and can predict success in the workplace and careers (Heckman & Kautz, 2012).

The world is a different place from 1945, when Bush published *Science: The Endless Frontier*. Even so, much of what he said in that report is still relevant in a modern context. However, it was impossible for Bush to fully envision how the advent of technology would enable communications and rapid flow of knowledge. We now understand that other factors, including global competence and soft skills, enhance STEM for societal benefit in the international community of nations in the twenty-first century.

Public Trust and Science Literacy in the United States

Even in 2015, the public doesn't trust scientists

Washington Post (Lynas, 2015)

Public trust in science is essential for both understanding and valuing science and STEM in modern society. In 2014, the Pew Research Center conducted a comprehensive survey of views on science in the United States (Funk & Rainie, 2015). Their survey is also relevant to a discussion about trust and confidence in the scientific enterprise. Their online survey polled two groups: (1) 2002 US adults, referred to as the "public" below; and (2) 3,748 US "scientists" who are members of AAAS (the American Association for the Advancement of Science). While STEM holds an important place in US modern society, with continued strong support from government funding for science, challenges remain. In an editorial in *Science*, US Senator Christopher Coons (2017) underscores the prevailing cloud of public distrust in science and the need for scientists to reach out to elected officials to better inform the legislative process.

Public respondents indicate an understanding of the overall benefit of science to society, with 79 percent agreeing that "science has made life easier for most people." Likewise, "a majority is positive about science's impact on the quality of health care, food and the environment" (Funk & Rainie, 2015: 1). About 70 percent of the public also say that government investments in science, engineering, and technology pay off in the long run. A majority of survey respondents, both among the public and scientists, also say that the scientific achievements in the United States are either above average, or the best in the world (Fig. 1.6).

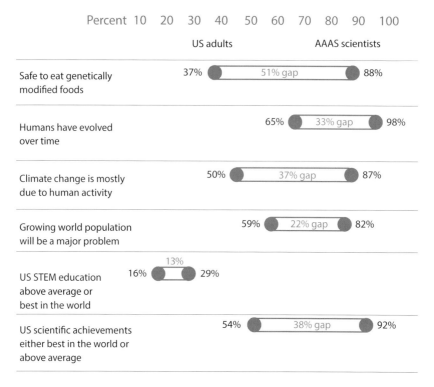

Fig. 1.6 Comparisons of Pew survey results from the public (US adults) and scientists (US members of AAAS) for questions pertaining to the perception of US science, controversial topics (climate change, evolution, genetically modified foods, and world population growth), and the perceived quality of STEM education in the United States (modified from Funk & Rainie, 2015).

Large gaps separate what scientists believe versus what the public believe. It likely will come as no surprise that for perceived controversial "hot-button topics" (Leshner, 2010) such as evolution, climate change, and genetically modified (Funk & Rainie, 2015) or genetically engineered (Frueh, 2017) foods, the gap is indeed profound. With regard to these topics, the Pew survey found significant differences between the public and scientists. Overall, 88 percent of scientists, but only 37 percent of the public, say that it is safe to eat genetically modified foods (Fig. 1.6). As we will also see in the following from a Gallup poll, the Pew survey (Fig. 1.6) indicates that far fewer of the public believe that humans have evolved over time (65 per cent) relative to scientists (98 percent). With regard to climate change, 87 percent of scientists, but only 50 percent of the public believe that climate change is due to human activity. Of additional relevance, 82 percent of scientists perceive the growing world population will be a major problem in the future, compared to only 59 percent of the public.

Why do such disparities exist between the public's and scientists' views of important topics of societal relevance? There is no simple answer to this question. Multiple factors are involved, including science literacy (i.e., understanding the context and processes of science), prior knowledge and belief systems, education, socioeconomic status, religion, and politics. Compounding this problem is the simple fact that science as a way of knowing has its own epistemology and jargon. Since their school days, the public had strived to know the truth or "the right answer." In contrast, scientists debate hypotheses, uncertainty, probability, and theory. The public seems to interpret this to mean that science is messy and lacks clarity, and they want to know the facts and answers, like the framework that they learned in school. Scientists are partially at fault with this disconnect. They oftentimes have difficulty communicating not just the process of science, but also its facts – and for the latter there is not always universal agreement. Likewise for controversial topics, the style of communication with the public has been ineffective in changing beliefs and attitudes (e.g., Nisbet & Scheufele, 2009). Take, for example, the fact that in the Pew survey 87 percent of scientists (not 100 percent) accept that climate change is mostly due to human activity. Climate deniers thus have the opportunity to say that not all scientists agree on this matter. Holt (2018) calls these general concerns "A tale of two cultures."

The next relevant concern is with current K–12 education. According to the Pew survey, although there is high confidence in the United States as a world leader in science for the common good, responses are far lower when asked about the efficacy of formal education. Only 16 percent of scientists and 29 percent of the public believe that K–12 STEM education is the best, or above average, relative to other industrialized nations (Funk & Rainie, 2015). According to another study, 90 percent of US adults believe that teachers play an important role in the well-being of our society, along with professors, business professionals, and lawyers. Despite this perceived value of teachers, their compensation is not commensurate with this status. In contrast, only 34 percent of US teachers feel that they are valued by society (NCEE, 2017). Low pay, lack of social status (in China teachers are regarded as similar to doctors), long hours, and other professional demands take their toll on the profession and affect student achievement. It is difficult to envision the United States as a world leader in STEM without an educational system that rewards teachers commensurate with their perceived value by society.

Hot-Button Topics, Politics, and Science

Similar to religion, in the long term it might have been good if the founders of the United States during the late eighteenth century had separated science and politics. However, science (certainly not STEM), oftentimes characterized as "natural philosophy" during that time (Johns, 2017), had yet to be recognized as a discipline of benefit to society. It is therefore understandable that politics have entered the domain of science and STEM, particularly in the post–World War II era of government funding, and in particular the founding of NSF in 1950. It likewise follows that the US Congress has oversight and expects accountability for its investment in NSF.

Fig. 1.7 A "hot-button" topic: evolution in US society. Top: acceptance of evolution in the 10 countries with the highest percentage of acceptance in comparison to the United States (Miller et al., 2006). Bottom: Gallup poll of beliefs related to humans, God, and evolution since 1982 (modified from Swift, 2017).

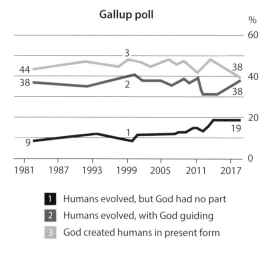

1 Humans evolved, but God had no part

2 Humans evolved, with God guiding

3 God created humans in present form

Although we will further discuss accountability in Chapter 2, it is also important for us to understand here the influence of science and politics.

The potential tensions between STEM, government, and even sometimes religion are indeed complex. The former executive director of AAAS, Alan Leshner, identifies "hot-button" STEM topics as ones that tend to polarize society (Leshner, 2010). Two hot-button topics of sustained relevance and debate include climate change and evolution. As a result of immense scientific evidence and the potentially deleterious effects on the Earth and its inhabitants, climate change, or more specifically global warming, has entered the world stage through multinational efforts such as the Paris Agreement (United Nations, 2017).

In the United States, evolution is also part of the social debate that involves both politics and religion. Miller et al. (2006) found that, of 34 countries, the United States ranks second to last in citizens' belief that evolution is "true" (Fig. 1.7) The authors of that study attribute this low acceptance to religious fundamentalism and the politicization of science. Of Similar relevance, a Gallup poll (Swift, 2017) of a sample of the US population indicates that 19 percent of respondents believe that "humans evolved, but God had no part" (in the process) (Fig. 1.7 (bottom), line 1); 38 percent believe that "humans evolved, with God guiding" (Fig. 1.7 (bottom), line 2); and 38 percent believe that "God created humans in [their] present form" (Fig. 1.7 (bottom), line 3). Thus, with the latter two percentages combined for the results reported in 2017, 76 percent of the respondents believe that God played some part in the origins and development of humans. Likewise, since 1982 the large majority of respondents in the Gallup poll have shown high acceptances of divine creation or intervention and this polling pattern has remained relatively stable during this interval. Scientists have not been successful in getting their message across to the US public.

It is indeed unfortunate that these hot-button topics in STEM have been elevated to national debate as part of elected officials' political platforms in the United States. With regard to evolution, for example, in the 2016 US presidential campaign, for the early group of Republican contenders only one explicitly stated that he believed in evolution. The other candidates mixed discussion of evolution with creation or intelligent design, or they were evasive or non-committal (Yuhas, 2015). We also see similar beliefs about the acceptance of human-induced climate change and its potential long-term economic impact (Hsiang et al., 2017), depending upon political affiliations of individual politicians. Despite the ongoing debate and current political polarization of science in the United States, 7 out of 10 people still believe that the effects of scientific research are more positive than negative for society (AAAS, 2018c).

Summary and Concluding Thoughts

This chapter is intended to provide the background and framework for the importance of science, and in a more inclusive context, STEM, in modern society. This discussion was largely influenced by the vision of scientists from the middle of the twentieth century, and, in particular, Vannevar Bush. Despite the fact that many of his recommendations still hold true today, it is perhaps not surprising the extent to which technology and globalization could not have been fully envisioned three-quarters of a century ago. We also explored the importance of, and tension surrounding, the notion of "basic science" and its importance in modern society. In addition, both old (evolution) and new (climate change) "hot-button topics" in science continue to be debated and further engrained within the political process. STEM continues to be an engine that drives successful economies, but it needs a well-trained and diverse workforce. We will revisit in greater detail many of the topics covered in this chapter later in this book. Case studies and examples, mostly from my professional experiences, will be presented to provide a better basis for how to develop NSF's Broader Impacts activities in the twenty-first century.

2 NSF and Broader Impacts

22 October 2014: Bad Day at the Office

When I arrived at the office and opened my email, I had an unexpected flurry of messages with both condolences and support for my NSF-funded FOSSIL (2017) project. Then US Senator Tom Coburn had just published his 2014 *Wastebook* highlighting excessive government spending by about 100 projects, and my FOSSIL project was one of those identified as wasteful spending from the National Science Foundation (NSF). It was called (Coburn, 2014: 52) "Facebook for Fossil Enthusiasts," which unfortunately misses the main point of the project. After an initial flurry of media attention, this matter has since mostly been forgotten, thankfully.

This dubious recognition harkens back to the days between 1975 and 1987 of US Senator Proxmire's "Golden Fleece Awards" that likewise took government-funded projects out of context and exposed them as examples of waste. Prior to 2014, I taught my students about Proxmire's Golden Fleece Awards and felt sorry for the scientists who were unfortunate enough to garner this notoriety. Now, I know how they felt, and it is embarrassing. It also further shows that as scientists we do a poor job of communicating to the public the importance of what we do and why they should care. NSF's concept of Broader Impacts grew out of this need for accountability and public understanding of science, and in a modern context, STEM.

Brief History of NSF and Development of Broader Impacts

Origin of NSF

Prior to World War II there was no NSF, and basic science was funded by a variety of sources, including industry and private philanthropy. It was clear that, with World War II raging in Europe, the United States would be drawn into conflict, and scientific research would have to help the war effort (e.g., Conant, 2002). During these years, US Senator Harley Kilgore tried on several occasions to sponsor legislation that would create NSF, but he did not succeed in moving this through the legislative process in Congress. As we learned in the previous chapter, at the request of President Roosevelt, Vannevar Bush's (1945) *Science: The Endless Frontier* forcefully described the importance of science and technology and its impact on the United States, both domestically and in the international arena. This report provided a vision and blueprint for NSF. In 1947 Congress passed the bill to establish NSF, but it was vetoed by President Harry S. Truman largely over matters of governance of this federal agency. After these concerns were resolved, Truman signed the bill into law on 10 May 1950, establishing the NSF "To promote the progress of science; to advance the national health, prosperity, and welfare; to secure the national defense, and for other purposes" (Fig. 2.1).

Much more will be said in the following about the development of Broader Impacts as part of NSF. It should be noted that this concept, although not branded specifically as such, was clearly embodied in the original intent of Congress, where it is stated that NSF would "appraise the impact of research upon industrial development and upon the general welfare" (Public Law 81–507, 1950). With regard to merit review, however, initially research grants were evaluated based "on the scientific merit of the suggested research including the competence of the investigator" (NSF, 1952: 51; also see Mazuzan, 1994). In his analysis of NSF's merit review criteria, Rothenberg (2010) states that external reviewers were also asked to evaluate proposals on other related characteristics, including uniqueness of the proposed research (in order to prevent duplication of effort), reasonableness of the budget, and quality of personnel and facilities at the host institution. Similarly, NSF program officers, who since the beginning have had considerable discretion with regard to funding recommendations, have been asked to evaluate proposals based on "national effort" (Rothenberg, 2010) as well as geographical and institutional distribution. Throughout its history, NSF and other members of the federal government have been concerned that funds should not favor "elite" institutions in a select number of states (also see EPSCoR, discussed later in this chapter).

The National Science Foundation has been governed by the 24-member National Science Board (NSB; Fig. 2.2) and an *ex officio* director, all of whom are presidential appointees. The first NSF Director was Alan T. Waterman, a former physics professor from Yale University and chief scientist at the Office of Naval Research. The first chairman of the Board of NSB was James B. Conant, president of Harvard University. The first-year appropriation from Congress was $225,000, largely for administrative and start-up costs (NSF, 1951). As of 1952, NSF's research divisions included: (1) Medical, Mathematical, Physical, and Engineering Sciences; (2) Biological Sciences; and (3) Scientific Personnel and Education (NSF, 1951). Medical research and Biological Sciences were merged soon thereafter (NSF, 1952).

In fiscal year 1952 – the first year that grants were awarded – NSF received a $3.5 million appropriation from Congress. This amount was an order of magnitude less than Bush (1945) had recommended. Nevertheless, during that year 96 research grants, 535 graduate (pre-doctoral) fellowships, and 38 postdoctoral fellowships were awarded. Budgets for research grants allowed for up to 15 percent indirect costs or overhead (which is quite different from what we see today, where indirect

[PUBLIC LAW 507-81ST CONGRESS]
[CHAPTER 171-2D SESSION]

[S. 247]

AN ACT

To promote the progress of science; to advance the national health, prosperity,
    and welfare; to secure the national defense; and for other purposes.

   *Be it enacted by the Senate and House of Representatives of the United
States of America in Congress assembled,* That this Act may be cited as the
"National Science Foundation Act of 1950".

ESTABLISHMENT OF NATIONAL SCIENCE FOUNDATION

SEC. 2. There is hereby established in the executive branch of the Government an
independent agency to be known as the National Science Foundation (hereinafter
referred to as the "Foundation"). The Foundation shall consist of a National
Science Board (hereinafter referred to as the "Board") and a Director.

FUNCTIONS OF THE FOUNDATION

SEC. 3. (a) The Foundation is authorized and directed--
       (1) to develop and encourage the pursuit of a national policy for the
promotion of basic research and education in the sciences;
       (2) to initiate and support basic scientific research in the
mathematical, physical, medical, biological, engineering, and other sciences, by
making contracts or other arrangements (including grants, loans, and other forms
of assistance) for the conduct of such basic scientific research and to appraise
the impact of research upon industrial development and upon the general welfare;
       (3) at the request of the Secretary of Defense, to initiate and support
specific scientific research activities in connection with matters relating to
the national defense by making contracts or other arrangements (including
grants, loans, and other forms of assistance) for the conduct of such scientific
research;
       (4) to award, as provided in section, 10, scholarships and graduate
fellowships in the mathematical, physical, medical, biological, engineering, and
other sciences;
       (5) to foster the interchange of scientific information among scientists
in the United States and foreign countries;
       (6) to evaluate scientific research programs undertaken by agencies of
the Federal Government, and to correlate the Foundation's scientific research
programs with those undertaken by individuals and by public and private research
groups;
       (7) to establish such special commissions as the Board may from time to
time deem necessary for the purposes of this Act;
and
       (8) to maintain a register of scientific and technical personnel and in
other ways provide a central clearinghouse for information covering all
scientific and technical personnel in the United States, including its
Territories and possessions.
       (b) In exercising the authority and discharging the functions referred to
in subsection (a) of this section, it shall be one of the objectives of the
Foundation to strengthen basic research and education in the sciences, including
independent research by individuals, throughout the United States, including its

Fig. 2.1 The first page of the National Science Foundation Act of 1950, Public Law
81-507 (1950).

costs are typically 50 percent or more in some budget categories). Also unlike today,
where proposals are highly structured with strict guidelines and submitted online,
during the first year of grant awards in 1952 there was no standard application form.
Proposers were advised to describe the proposed research, methods (procedures),
facilities, personnel, and budget (NSF, 1952).

Fig. 2.2 National Science Board, July 1951. The chairman of the Board, James B. Conant, is in the front row, center. The first NSF director, Alan T. Waterman, an *ex officio* NSB member, is in the front row to the right of Conant in this photo (NSF photo; NSF, 2017a).

Waterman estimated that at least 40 percent of the proposals received during that first year were fundable, but only $1.1 million could be allocated from the $13 million requested in the proposals. The average research grant was for $11,156 and funded for 1.9 years. These grants were distributed across 59 institutions from 33 states and the District of Columbia and Hawaii. The number of proposals submitted versus the 96 funded is not listed; therefore the funding success rate (percentage) cannot be determined. In the 1952 report (NSF, 1952) the statement was made that: "It is clear, however, that limited Foundation funds for research support has discouraged many competent investigators from submitting proposals." This will likely come as no surprise to those investigators working in the twenty-first century.

During the first decade of NSF, some emphasis was placed on the importance of funding for "big science," that is the IGY (International Geophysical Year, 1957–1958; Inset 2.1). Likewise, during this time, there was a substantial increase in NSF's budget in response to the Soviet Sputnik program and the "Space Race," in which there was a competitive spirit to be first into space (Abramson, 2007; Mazuzan, 1994). The successful development of the vision, policies, and procedures that occurred during NSF's beginning set the stage for continued growth of this agency thereafter.

Inset 2.1

The virtues of the graduate student and postdoctoral support have been realized since the days of Bush (1945). Of local relevance, the PhD conferred to Dr. Elizabeth Wing (Fig. 2.3) from the University of Florida was supported by an NSF Fellowship (Wing, 1962). Liz also received support for several other projects that

Fig. 2.3 Dr. Elizabeth Wing, Curator Emerita, Florida Museum of Natural History, University of Florida, in the late 1970s. Starting as a graduate student in the 1950s, Liz was supported by NSF funding for her research in the emerging field of zooarchaeology. For this work she was elected to the National Academy of Sciences in 2007. The skeleton of the horse *Equus* in the foreground is part of the comparative osteology collection in zooarchaeology (FLMNH archives photo).

led to the development of the new science of zooarchaeology (Wing, 2005), in which diets of past human cultures are reconstructed by analyzing associated fauna. For her contributions to this field, and starting from within the male-dominated world of science in the 1960s, Liz was elected to the National Academy of Sciences in 2007.

The Golden Age

Ten years after the first grants were awarded, NSF had increased its budget by two orders of magnitude, from $3.5 million in 1952 to $263 million in 1962 (NSF, 1952; 1962). It also had greatly expanded its programs to include more emphasis on science education (e.g., adding undergraduates and K–12 teachers), international activities (e.g., for topical symposia and NSF's first foreign office opened in Tokyo), and facilities support (the Florida Museum of Natural History building [Fig. 2.4] was funded by NSF during this time; Dickinson, 1965; 1966).

The 1960s also saw continued increase in annual budget appropriations from Congress (Fig. 2.5). At the same time, the percentage of proposals funded rose to a peak of about 60 percent in the late 1960s (Rothenberg, 2010), a level of success that has not been seen since that time (NSF's average funding rate for the period 2001–16 was 27 percent; compiled from NSF, 2018a). The National Science Foundation thus fared well

Fig. 2.4 Florida Museum of Natural History, Dickinson Hall, completed in 1970 and located on the central campus of the University of Florida. The construction of this building was funded by a $1.1 million grant from NSF, plus state and private funds (FLMNH archives photo).

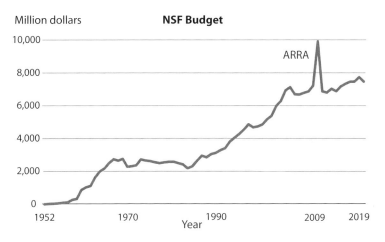

Fig. 2.5 The NSF budget from 1951 to 2019 in real dollars allocated (not adjusted for inflation). The large increase in 2009 represents supplemental funding provided by the ARRA (American Recovery and Reinvestment Act). Sources: NSF (1951–2015; 2016a; 2017b; 2018a; 2018b).

during the administrations of presidents Kennedy and Johnson. Mazuzan (1994) referred to this time as the "Golden Age." He also noted that, likely resulting from a modest upbringing and a graduate of the non-elitist Southwest Teachers College in Texas, President Johnson was concerned about democratizing support for science, that is emphasizing the concept of spreading funds across diverse institutions and states

(Mazuzan, 1994). Johnson's point of view heralded NSF's EPSCoR (Experimental Program to Stimulate Competitive Research), which has continued for decades.

President Kennedy issued an executive directive that created the Office of Science and Technology in 1962, which would be led by the presidential science advisor. Emphasis continued to be placed on accountability and societal impact. In 1963 Kennedy addressed the one-hundredth anniversary of the National Academy of Sciences and noted that "scientists alone can establish the objectives of their research, but society, in extending support to science, must take account of its own needs." Along with the president's brother, Senator Edward Kennedy, Representative Emilio Daddario sponsored an amendment that broadened the scope of NSF's activities while at the same time requiring annual review and oversight of this agency by both Senate and House science committees.

Vietnam and "Golden Fleece" Eras

Mazuzan (1994) asserts that, for a variety of reasons, the period encompassing the late 1960s and early 1970s was a tumultuous time for NSF, and basic science in general. This related to the impact of the Vietnam War, various entitlement efforts, tensions between President Richard M. Nixon and his science advisors, and a shaky transition to the new NSF director, William D. McElroy. Despite these challenges, however, congressional appropriations for NSF continued to increase during the height of the Vietnam War in the late 1960s (Fig. 2.5).

Despite its original mission to promote basic science, there was growing interest in developing applied research within the purview of NSF. Stemming from the Daddario–Kennedy amendment, the RANN (Research Applied to National Needs) program gave NSF the authority to conduct applied research – that is, research that has immediate societal benefit. This initiative potentially broadened the mission of the agency and thus, more directly, addressed socially relevant science. To quote Mazuzan (1994: np), RANN "blossomed and then faded during the period 1969–1977." Also, likely stemming from the Daddario–Kennedy amendment, the Technology Assessment Act of 1972 (Public Law 92-484, 1972) authorized NSF to support activities related to the "effects of scientific application upon society." In response to an increasing number of proposals, in the early 1970s, the NSB attempted to clarify the selection criteria for proposals (Rothenberg, 2010). Thus, in 1974 NSF implemented 11 new selection criteria for their research projects, which were grouped into four broad categories designated A–D (NSF, 1974), paraphrased here as the (A) competence of the researcher and his/her institution; (B) value of the science proposed; (C) utility or relevance, including applications; and (D) effect in scientific infrastructure of the United States. With the exception of A, which was used to evaluate all proposals, criteria B through D were used to different degrees depending upon the project (Rothenberg, 2010). Although implicit, neither the words "broader" nor "impact(s)" were used in the description of these criteria (NSF, 1974). Additional tweaking of the review criteria occurred throughout the 1970s and 1980s, but these changes were largely guided by the four categories listed above. In 1986, NSF specifically utilized the term "merit review," which remains fundamental to its proposal evaluation process to the present day (Rothenberg 2010).

A controversy rocked NSF in the 1970s: A course for fifth-graders called "Man: A Course of Study" (MACOS) was funded by NSF and largely created by the non-profit Educational Development Corporation. This curriculum was subsequently implemented widely across the United States. The MACOS program was purported by its critics to distort family values, including beliefs and morality, and was targeted by fundamentalist religious groups. Interestingly, MACOS emphasized the process of science along with its content and "spiral curriculum," in which science learning concepts are taught with increasing complexity as students move up through grades. In certain respects, this framework is similar to the concept of learning progression, which is fundamental to K–12 standards in the twenty-first century (e.g., NGSS, 2013). Although MACOS faded away, a documentary film about the controversy surrounding this project was released in 2004 (Kincheloe & Horn, 2007).

The notion of equal geographic distribution of support, and an aversion to favoring "elite" institutions, has been fundamental to NSF's mission since its early days. In 1978 the NSB approved EPSCoR. This effort, which continues to the present-day, has the goal of helping to support research and education in states that have relatively low levels of NSF funding, in order to improve their research competitiveness (NSF, 2017c). A state is eligible to participate in the EPSCoR program if its most recent three-year level of NSF research support is equal to or less than 0.75 percent of the total NSF Research and Related Activities (R&RA) budget (NSF, 2017c).

NSF Golden Fleece Awards		
1975	$85K	Why people fall in love
1976	$46K	Scantily clad women & Chicago taxi drivers
1976	$397K	Consumer legislation
1977	$N.A.	Use of taxpayer money
1979	$39K	Himalayan...Sherpas
1981	$144K	Pigeon supply...demand
1987	$10K	Bullfights & Spain

Inset 2.2

During the late 1970s, US Senator William Proxmire took aim at many examples of wasteful government spending by federal agencies. The NSF was not spared in his monthly press releases, nor in his announcements on the floor of Congress. During the interval between 1975 and 1987 Proxmire "awarded" 168 Golden Fleece Awards, seven of which involved NSF grants (Proxmire, 1975–1987; Inset 2.2). As indicated in the titles, Proxmire had the social and behavioral sciences in his sights. His dogged, no-nonsense style in pursuit of wasteful spending received considerable recognition. The *Washington Post* referred to these awards as "the most successful public relations device in politics today" (Mills, 1988). Proxmire's Golden Fleece Awards presage similar initiatives in more recent years, including US Senator Coburn's (2014) *Wastebook*.

The "New" Review Criteria

In 1997, the NSB approved the two current merit review criteria – Intellectual Merit and Broader Impacts. In what has been seen by some as a departure from previous policy, these two criteria were intended to have equal weight in proposal decisions. The addition of Broader Impacts has been decried by the research community as a new wrinkle – and to some a thorn in the side of proposers. Despite these protestations, the original act that created NSF, Public Law 81-507 (1950; Fig. 2.1), implies the concept of Broader Impacts and societal benefit (e.g., "impact of research … upon the general welfare") that has been at the core of NSF since its beginning. In 1997, NSB chair, Richard Zare, stated that the "new" criteria did not represent any real change – in fact, they simplified the ability to judge proposals (Rothenberg, 2010).

   For the next decade and a half, NSF's Broader Impacts criteria had a checkered history and one of dubious acceptance by scientists (e.g., Sarewitz, 2011). Frodeman et al. (2013: 153) noted that "*broader impacts* meant an additional burden in the lives of increasingly harried researchers." Thus, for a variety of reasons, also including reviewers' personal opinions about the relative importance of Broader Impacts or their apparent lack of understanding of them, these criteria were likely inconsistently applied during the review process (Brainard, 2008). At that time, and likely continuing today in some circles, many members of the scientific community equate Broader Impacts with outreach, the latter of which is considered of dubious importance to those researchers focused on scientific discoveries (Mervis, 2007). A basic and pragmatic concern voiced by scientists has been that they neither knew exactly what Broader Impacts are, nor how to do them. In fact, NSF realized this problem and stated that:

> While most researchers know what is meant by Intellectual Merit, experience shows that many researchers have a less than clear understanding of the meaning of Broader Impacts.
>
> (NSF, 2007)

In an important Dear Colleague Letter (DCL) (NSF, 2007), NSF reiterated the importance of Broader Impacts and also described five kinds of representative and appropriate activities (Box 2.1). Until recently (NSF, 2016b) I have used the 2007 DCL when teaching about Broader Impacts because it is a tangible set of activities that make sense. However, this list has been judged by some members of the basic science community interested in NSF funding as being too prescriptive (Hand, 2011). Rather than being viewed as suggestive activities, some scientists interpreted these to represent a set of rules to be carefully followed. Thus, some proposers believed that they needed to do all of these activities to have a successful Broader Impacts plan. Mardis et al. (2012) analyzed a set of NSF proposals in the NSDL (National Science Digital Library) program. They found that some funded proposals describe Broader Impacts plans that are aspirational and resulted in incomplete activities at the time the project ended.

Box 2.1 NSF's Broader Impacts: Recent Updates

Dear Colleague Letter (NSF, 2007): Representative Activities

1. Advance discovery and understanding while promoting teaching, training, and learning
2. Broaden participation of underrepresented groups
3. Enhance infrastructure for research and education
4. Broaden dissemination to enhance scientific and technological understanding
5. Benefits to society

America COMPETES Act (2010)

Sec. 526: Broader Impacts Review Criteria:

1. Increased economic competitiveness of the United States
2. Development of a globally competitive workforce
3. Increased participation of women and underrepresented minorities in STEM
4. Increased partnerships between academia and industry
5. Improved pre-K–12 STEM education and teacher development
6. Improved undergraduate STEM education
7. Increased public scientific literacy
8. Increased national security

Proposal & Award Policies & Procedures Guide (NSF Solicitation 17-1 in NSF, 2016b)

Broader impact outcomes include, but are not limited to:

1. Full participation of women, persons with disabilities, and underrepresented minorities in STEM;
2. improved STEM education and educator development at any level;
3. increased public scientific literacy and public engagement with science and technology;
4. improved well-being of individuals in society;
5. development of a diverse, globally competitive STEM workforce;
6. increased partnerships between academia, industry, and others;
7. improved national security;
8. increased economic competitiveness of the US; and
9. enhanced infrastructure for research and education.

2010 and Onward: Broader Impacts 2.0

In 2010 Congress enacted the America COMPETES Act with the goal of making the United States more competitive in STEM worldwide. This act pertained to several federally funded agencies, including NSF. It more clearly articulates activities that are considered important, not just for basic research, but also for diversity, broadening representation, academia and industry partnerships, K–16 STEM education, as well as the importance of scientific literacy and increased national security. These essential elements were then codified in NSF's annual guide to proposers soon thereafter (NSF, 2012a). A section specifically describing Broader Impacts had become a required part of each proposal. Furthermore, the merit review criteria expected reviewers to address "equally" Intellectual Merit and Broader Impacts. Frodeman et al. (2013) have called these changes "Broader Impacts 2.0." In the inevitable and incremental changes that occur to the details of Broader Impacts, nine intended outcomes of Broader Impacts are described (Box 2.1) in the recent Proposal & Award Policies & Procedures Guide (NSF, 2016b). Similar to the 2007 Dear Colleague Letter, these may be interpreted to be too prescriptive; however, they provide clear direction of the kinds of Broader Impacts that are envisioned.

The past decade or so has also seen research offices at some universities provide professional development and administrative personnel in support of faculty who desire to better develop Broader Impacts. In addition, in 2013 a professional organization, now called the National Alliance for Broader Impacts, or NABI (NABI, 2017), was created to provide an online community of practice (*sensu* Wenger et al., 2002) for researchers, early-career professionals (e.g., graduate students), and administrators interested in better understanding the background and best practices of Broader Impacts. Since 2013 this organization has also hosted an annual meeting that includes sessions, themes, and topics relevant to Broader Impacts. There has been a recent move to start a peer-reviewed journal devoted to studies and examples of Broader Impacts (NABI, 2017). The positive benefits of Broader Impacts therefore appear to be gaining ground within the culture of academia.

Summary and Concluding Comments

Challenges to NSF: Anonymous Peer Review

This chapter was presented to better understand and frame the importance and impact of basic scientific – and in the twenty-first century, STEM – research for the ultimate benefit of society. The importance of Broader Impacts and effective communication of science to the public has also been underscored by challenges to NSF projects by Proxmire's Golden Fleece Awards in the 1970s and 1980s and Coburn's (2014) more recent *Wastebook*. More recently a new kind of government scrutiny has emerged, one that challenges the proposal peer-review process that has been a core principle of NSF.

Since its early days, NSF has sought input from experts who evaluate proposals. This process has been anonymous, meaning that only NSF staff know who is reviewing proposals. Likewise, the internal documentation of NSF's decisions to fund a proposal, or not, is also confidential. This has resulted in a process that is deemed both effective and equable. Proposers never learn through NSF who has reviewed their proposals; this protects the identity of the reviewer and the integrity of the peer-review process. Over the past decade, US Representative Lamar Smith, chair of the House Committee on Science, Space, and Technology, has requested copies of confidential documentation of selected NSF proposals. Initially in 2014, 20 funded projects were chosen that were deemed frivolous. Under public records laws, the request was made for the confidential documentation to be provided to Congress. Because it could potentially compromise the anonymous peer review (e.g., if some identities were "leaked"), NSF pushed back and allowed congressional staffers to study redacted versions of the requested proposals. Likewise, the staffers were not allowed to photocopy or otherwise reproduce the relevant information. Mervis (2014) notes that:

> Federal lawmakers have had a long tradition of taking potshots at grants they don't like, but Smith's April 7 request for material on the 20 grants was unprecedented, and it created a major dilemma for NSF.

Since the original request, dozens of other proposals have been solicited. Nearly two-thirds of these proposals were funded through NSF's often-maligned SBE (Social, Behavioral, and Economic Sciences) Directorate (Mervis, 2014). This harkens back nearly a half-century to Proxmire's Golden Fleece Awards that singled out projects funded through these and similar programs. More recently, Smith introduced legislation that potentially would have changed the integrity and confidentiality of NSF's peer-review system. That bill was not passed in its original form. Nevertheless, Smith and like-minded federal legislators continue to push for research funded by NSF that is of increased relevancy and "in the national interest" (Mervis, 2017a). According to Mervis (2017b), Smith retired from Congress in 2018, and as such it remains to be seen if this is "the first step towards a more civil science committee in the U. S. House of Representatives."

Government scrutiny of NSF is not confined to the House. In 2017, US Senator Rand Paul proposed legislation that likewise would have made the proposal review process more transparent. Proposal review panels would include members with no research expertise in order to determine if the projects being funded "deliver value to the taxpayer" and represented "silly research." Another US Senator, Gary Peters, responded that "silly-sounding titles [can mask] the true scientific merit and potential broader impacts of the work" (AAAS, 2017; Braaten, 2017). These conversations likewise harken back to the days of Proxmire and underscore the importance of effective communication of the value of science to the public, including decision-makers in the US Congress.

Stability of NSF and Constancy of Broader Impacts

An analysis of NSF's history teaches us that the core mission and goals of this federal agency have remained much the same since the 1950s. The original intent of NSF was to promote basic scientific research, but with an understanding that these pursuits should be of benefit to society and the national interest. Despite claims that Broader Impacts were newly foisted upon the community as an official merit review criterion in 1997, although not named as such, this concept has been part of NSF since its beginning in 1950. Analysis of the evolution of NSF's review criteria during the past three-quarters of a century also indicates consistent emphasis on societal impact and benefit.

There have been several times since 1950 when influences external to NSF have advocated for a more direct relevance, application and translation (e.g., the analog of translational medicine) for societal benefit. This resulted in experimentation with the direct application (e.g., RANN) of science and technology. These programs, however, were usually initiated because of external (i.e., legislative) influences and potentially would have diluted NSF's mission and moved it away from basic science, as Bush (1945) had envisioned. In so doing, it would have joined the ranks of about two dozen other federal agencies that sponsor research with specific, directed outcomes (Grants.gov., 2017). Related to NIH (the National Institutes of Health), NSF originally had medical research within its purview, but that emphasis was subsequently dropped from its mission.

Certain landmark events and actions have further underscored and elevated the importance of Broader Impacts and its precursors during this time. These have included, for example, other governmental initiatives such as the America COMPETES Act of 2010 and more recent guidelines to proposers of NSF grants (e.g., NSF, 2016b). If we learn anything from history, we can see that Broader Impacts have received continuous tweaking, as well as periodic, major revisions. These changes typically pertain to increased emphasis on clarification of expectations for Broader Impacts plans and activities. At the same time, these changes can reflect the increased accountability within the prevailing federal and societal attitudes towards basic science and STEM. History teaches us that Broader Impacts will not go away. They will continue to remain part of the fabric, culture, and mission of NSF and therefore promote the contributions of science and STEM in twenty-first-century society.

3 Innovation, Opportunity, and Integration

A "Once-in-a-Century Opportunity"

In January 2002 I was part of a team from our museum that traveled to Panama to visit the Smithsonian Tropical Research Institute (STRI). We toured scientific field sites and experienced the local biodiversity around the Panama Canal. For that week we were the guests of a philanthropist from our home town (Gainesville) who financed the trip. We flew to Panama in a chartered jet, stayed at fine hotels, and dined at the best eating restaurants that Panama had to offer. This was a far cry from the economy flights, mediocre hotels, and low-budget field work that I had typically experienced as a scientist. When I was high above Metropolitan Park in STRI's research crane, it was a thrill to see the jungle canopy and its immense biodiversity. Despite my aversion to hot, humid field work, something about Panama captured my imagination. I decided that I wanted to start a paleontological research project there. I knew that in the 1960s Smithsonian and Canal Zone scientists had collected 20 million-year-old (early Miocene) horses, rhinoceroses, extinct deer-like mammals, and rodents from sedimentary exposures along the Panama Canal (Whitmore & Stewart, 1965). My research deals with these kinds of fossil mammals, and I wanted to learn more about them and the history of the rainforest in what is now Panama.

I returned from Panama and strategized about how to begin a paleontological field program there and expand the work that had been done some 40 years earlier. After receiving a small seed grant from the University of Florida (UF), I submitted three successive proposals to NSF; despite solid reviews, these did not distinguish themselves from the other deserving projects. All of these proposals were declined until 2007, when I received a small NSF research grant to work in Panama. During this time I learned that Panama was planning to undertake a major expansion of the Canal. These excavations would uncover the sediments that I was studying with my small NSF grant. The proposed excavations were on a scale not seen since the original excavation a century before (the Canal opened in 1914; McCullough,

1977). At the same time, NSF (2009a) issued a solicitation for their PIRE program (Partnerships for International Research and Education) that seemed like a way to fund our growing field project in Panama. PIRE projects are extremely competitive, with hundreds of preproposals submitted that were winnowed down to fewer than 100, from which a dozen were selected for funding. Many of the other proposed projects likely had strong research agendas. We needed something to distinguish the Panama project from the rest of the pack. We thus marketed the project as a "once-in-a-century opportunity" to advance understanding of the ancient New World topics resulting from the new excavations along the Canal. This project was funded and ran its course from 2010 until 2016 (Panama PIRE, 2017), roughly coinciding with the excavations and then opening of the new portions of the Canal. One colleague remarked to me "they had to fund the project given the special opportunity." While these were kind words, and hardly a mandate by itself, marketing the scientific opportunity seemed to add appeal to our Panama project.

Introduction

Several of the chapters here could be entire books in themselves, and this is one of them. The topics chosen – innovation, opportunity, and integration – are potentially diffuse and all-encompassing at the same time. These are intangible aspects to successful NSF projects that are hard to realize, predict, describe, or evaluate. Scientists should challenge themselves to look for these aspects from the beginning of their envisioning and development of NSF projects. The focus here will be on past experiences, representative examples, and interesting case studies.

Innovation

> Transformative research (TR) statements in scientific grant proposals have become mainstream … it is rarely possible to predict the transformative nature of research.
>
> Gravem et al. (2017: 825)

When Broader Impacts were first presented to the research community, scientists scrambled to figure out what they should do. For some, the easiest idea that still fell within their comfort zone of academia was to propose that they would infuse the content of their research into the courses that they teach, and thus spread pearls of wisdom to the masses. This strategy soon became passé, and other activities and plans were required in order to fulfill a more competitive Broader Impacts plan.

In order to stay current and sustain success and relevance, science and STEM must continuously improve and make discoveries that "advance the field." In a word, they must be innovative. Nestled within the word innovative is the Latin root, "nova," meaning new or novel. To do what has been done before is not innovative, or may represent "incremental" science – those activities that add limited new knowledge or content to a discipline. The National Science Foundation and other organizations such as private foundations have the expectation that the projects that they fund are innovative. The bar is high – as it should be; taxpayers and philanthropists should expect innovations and major breakthroughs from their investments. Many nouns and adjectives express the concept of innovation (Inset 3.1); all of them fit within the theme of advancing science and STEM in the twenty-first century for the overall benefit of society.

> **Innovation**
>
> *Synonyms, or similar words*
>
> Breakthrough
> Clever
> Cutting edge
> Creativity
> Imagination
> Improvement
> Inspiration
> Novel, new
> Progress
> State-of-the-art
> Transformation

Inset 3.1

Words such as "cutting edge" and the like are oftentimes associated with innovation. Of all of these words, *transformation*, and its adjective *transformative* are important, because the latter is specifically used in some NSF solicitations as an expectation, particularly in larger, complex, and more expensive programs. For example, the STC (Science and Technology Centers) program, which a colleague referred to as "the crown jewel of NSF," raises the bar on transformation. Thus, the STC program "supports innovative, potentially transformative, complex research and education projects that require large-scale, long-term awards" (NSF, 2014a).

The challenge with calling a project transformative is that, although it is frequently expected, it is difficult to quantify or predict (Gravem et al., 2017). For reviewers, it is therefore oftentimes a challenge to judge a project as being innovative, or transformative. In 2016, the National Academy of Sciences (2017a) convened a workshop entitled "Advancing Concepts and Models of Innovative Activity and STI Indicator Systems." This high-level workshop brought together a distinguished group of international scholars and practitioners involved in academic research, the private and business sector, and public policy. The goals of this workshop included a better understanding of metrics that could be used to quantify innovation in a nation's R&D (research and development). As it turned out, although STEM was part of the discussion, most of the tangible aspects of this workshop focused on innovation metrics in applied research and technology transfer over the past few decades. Building on other reports (e.g., the Oslo Study; OECD-Eurostat, 2005), it is clear that any definition of innovation must include an understanding of implementation and how this novelty can be described in terms of metrics and outputs. These can include, for example, number of patents, tech startups, increase in the workforce, and the like (*Science*, 2018). The problem with applying these parameters to the NSF proposal process is that the focus on quantifying innovation so far has been on implementation and outputs, and thus the application of research, or an a posteriori view of research investment. This does not help the evaluation or quantification of the specific proposal or innovation when it is proposed a priori during the review

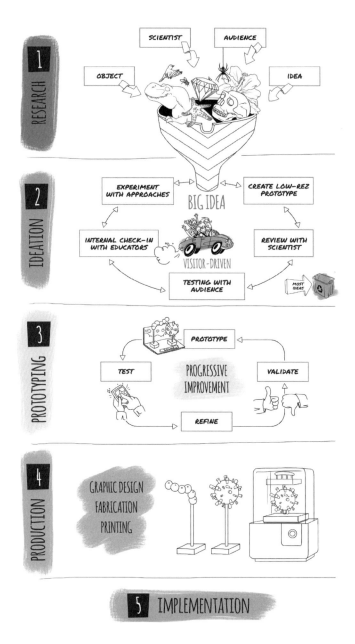

Fig. 3.1 An example of "ideation" during exhibits and project development workflow (Bolton et al., 2018; Office of Education and Outreach, National Museum of Natural History, Smithsonian Institution). Ideation is the process of brainstorming to create ideas, concepts, and/or plans.

process. Perhaps a good middle ground during the NSF review process is to evaluate the *potential* for innovation implementation using deliverables, outputs, and impacts described in the specific proposal(s) under consideration.

Brainstorming and "Ideation"

It is well known that, although innovation sometimes happens in a vacuum by individuals, more often it is best for this process to unfold through collaboration (Paulus & Nijstad, 2003). Consider the concept of "think tanks" in which ideas are generated through iterative and strategic brainstorming. Brainstorming as an idea generator is anything but haphazard, and best practices have been developed since this term was invented in the middle of the twentieth century (Markman, 2017). Bolton et al. (2018) present the innovation workflow process (Fig. 3.1) in which they develop exhibits and other new programs and activities at the Smithsonian Institution. This process follows a series of steps, including the "ideation" phase in which ideas are produced via collaboration within the project team. As with all such processes intended to result in innovation, trial and error, prototyping, and iterative improvement are key components to potentially successful outcomes.

Case Examples of Innovation

Curate My Community

Many innovative projects could be highlighted here. I saw one memorable example in the Cincinnati, Ohio, airport when I traveled to a meeting. As I exited my flight, several fossil skeletons were on display in the airport, and these were united by the theme "Curate My Community." Natural history specimens and displays are not new to airports (e.g., Hoganson, 2011). They can serve a valuable function to potentially engage and attract the public to the local museum. This exhibit provides a particularly innovative solution to a problem.

   Curate My Community was undertaken by the Cincinnati Museum Center. This project is innovative because, in addition to what had been done before at other airports, the museum staff used this opportunity to solve a problem in a particularly imaginative way. The problem was that the museum was undergoing a multi-year major building renovation project, one that would result in closing the exhibits to public view. A related issue was that the then-existing exhibits needed to be completely gutted and the objects on display needed to find a temporary home elsewhere. These objects included vintage automobiles and airplanes, a taxidermied polar bear, and numerous fossil skeletons of dinosaurs and extinct mammals such as giant sloths and mastodons (Fig. 3.2). Rather than put these 700 iconic objects in "dead storage," the museum developed the Curate My Community initiative. They networked with community partners to loan the specimens for display in more than 50 local buildings, including libraries, universities, breweries, shopping centers, a children's theater, and, as I experienced,

Fig. 3.2 Skeleton of an American mastodon (*Mammut americanum*) on display at the Cincinnati/Northern Kentucky Airport as part of the Curate My Community program sponsored by the Cincinnati Museum Center (Rosemary Pennington photo).

airports (Lima & Linzinger, 2017; Fig. 3.2). Interpretive signage not only described the object on display, but also branded the unifying theme of Curate My Community. After the building renovations were completed, the artifacts and iconic specimens were brought back into the museum. Thus, the Cincinnati Museum Center concluded an unprecedented and innovative opportunity to impact the community while at the same time solving the space and storage problem. As further validation of its innovation, this project recently won the Roy L. Shafer Leading Edge Award from ASTC (the Association of Science and Technology Centers). These awards are given "in recognition of extraordinary accomplishments in visitor experience, business practice, and leadership in the field that not only enhance the performance of their own institutions but also significantly advance the mission of science centers and museums" (ASTC, 2018).

Pop Culture Audiences

As we will learn in the following, scientists wanting to achieve societal impact oftentimes will do this by outreach to various kinds of audiences, such as K–12 teachers and students, museums, and festivals. While the formal K–12 teachers and students represent a captive audience (they have to be there), those that visit museums and festivals tend to be self-selected – they are generally predisposed toward an interest in science and STEM. We sometimes try to stretch the boundaries by doing themed displays (pop-ups) at farmers markets and other kinds of public gatherings that are not meant for science learning.

Santos et al. (2018) report on an innovative program to bring the science of paleontology to pop-culture festivals, in this case "comic-cons." They capitalize

Fig. 3.3 Paleontology, pop culture, and Pokémon. An exhibit display at a comic-con festival in Los Angeles, California. In one activity, the paleontology content of fossils (mastodon skull) in the right front is compared to fictional Pokémon characters behind (Gabriel-Philip Santos photo).

on public interest in fossils and develop themed exhibits that relate fossil animals (e. g., mastodons) to fictional Pokémon characters (Fig. 3.3). While this seems way out on the fringe, it provides a common ground to integrate public interest in paleontology and pop culture, and in so doing promote science learning. It therefore reaches an audience that might not otherwise be attracted to other venues involving science content.

Risk and ROI

No discussion of innovation is complete without a consideration of risk and ROI (return on investment). As an organization, NSF is careful about exposure to too much risk ("risk tolerance"); this is good fiscal practice and ensures, to the extent possible, maximum ROI. This makes sense, given the source of NSF funding and both congressional oversight and taxpayer accountability. Nevertheless, innovation requires risk in order to optimize new discoveries that advance scientific knowledge. The National Science Foundation manages risk in several ways. Many of the highest risk programs, like SGER (Small

Grants for Exploratory Research) encourage high-risk research with the potential for breakthroughs (Gravem et al., 2017), thus resulting in potentially high reward. Likewise, many programs agency-wide have smaller "seed grant" programs to get projects started, determine feasibility, and provide "proof-of-concept," which are in a sense risk-management strategies. All of these, however, are relatively low-budget programs (i.e., tens to hundreds of thousand dollars). The Panama PIRE project described above was preceded in 2006 by a small ($\sim$$10,000) planning grant awarded from NSF's Office of International Science and Engineering (OISE) to travel to Panama to build partnerships. The results of this planning grant demonstrated the feasibility of the partnerships that were then elaborated upon with the larger, multi-year grant.

Although not called as such, NSF's peer-review system has had some risk-management strategies in place for decades. Part of the Intellectual Merit questions asked of reviewers relate to risk management, including the quality of the research team, feasibility of the project, and capacity of the host institution(s). In addition, a strong predictor of productivity (but not success, per se), relevance, and capacity is contained within the "Results of Prior NSF Support" section that is required in the Proposal Description.

NSF's implied risk management is inherently conservative. With the exception of smaller grants, such as for exploratory research and seed money, the larger "standard" or core program grants are subject to both individual peer-review and panel deliberations. Despite what may be desired from the process, reviewers tend to be risk averse and not particularly savvy or supportive of innovation that does not lead to appropriate ROI. In certain instances, NSF program officers can "override" reviewer feedback and ratings if they see innovation potential, even if there is risk involved. This, however, is an ad hoc system. Thus, there is a paradox: larger projects are expected to be innovative and transformative, but the review process does not allow much latitude for risk. Without risk, most NSF projects largely represent safe science with only some latitude built in for risky innovation and experimentation.

Return on investment is a term adapted from the business and economic sectors of society (also see Chapter 17). The concept of ROI implies tangible metrics, deliverables, and product development that typically lie within translation of bench science for the economic benefit of society. But traditionally ROI has not been applied within the sphere of basic research. More recently, some of the programs within NSF's portfolio, including Small Business Innovation Research (SBIR), have started to accept "innovative proposals that show promise of commercial and societal impact in almost all areas of technology" (NSF, 2017d: 1). These kinds of programs are intended to increase ROI in the applied sphere of research. In addition, a case can be made for NSF to consider the benefits of ROI for free and open-source software (Pearce, 2016). Other than short-term metrics, ROI applied to strategic initiatives such as broadening participation, development of the STEM workforce, and overall societal benefit accrued from funded research will be more elusive. Regardless of this challenge, we will likely see more emphasis in the future on accountability and quantification of basic research ROI resulting from NSF-funded projects.

Opportunity

Scientists searching for NSF funding should not underestimate the value of finding special opportunities and tailoring their research to specific programs that might not previously have been on their radar or within their comfort zone. Such was the case with the Panama PIRE (2017) project described earlier. During a meeting with an NSF program officer in 2012 about the Panama project and my emerging interest in working with STEM teachers, she pointed me in the direction of the RET (Research Experiences for Teachers) program in her portfolio. While most researchers have heard of REUs (Research Experiences for Undergraduates), RETs are one of the best-kept secrets at NSF and are undersubscribed in some divisions. For a variety of reasons, RETs and REUs are ready-made for a Broader Impacts plan of activities and can (and should) be integral to the proposed research. Our "spin-offs" of the Panama PIRE included supplements and a stand-alone RET "site grant" (GABI RET, 2017). Thus, during a five-year period (2012–16) we engaged more than 50 teachers to collect fossils with us around the Panama Canal. In so doing, we were better able to understand the biodiversity that lived in the ancient New World tropics over the past 20 million years. Had we not availed ourselves of the RET opportunity, our teacher–scientist partnerships would not have been developed over the past several years.

The National Science Foundation has a range of opportunities that perhaps in the views of business-as-usual investigators are outside their comfort zones. But many funding opportunities exist if one keeps an open mind and is willing to be flexible. These can include supplements to existing grants, or new opportunities oftentimes announced through additional solicitations (e.g., for REU/RET sites; NSF, 2017e) or Dear Colleague Letters (e.g., for public participation in scientific research; NSF, 2017f).

The National Science Foundation is an organization that is constantly changing in response to factors such as input from the National Science Board (NSB) as well as perceived trends within science and STEM research. Since 2010, themes and new programs have been developed at NSF that have addressed, for example, climate change, sustainability, data science ("Big Data"), and citizen (community) science. The National Science Foundation also seeks community input and then forms internal working groups to develop new programs and solicitations. Each of these, and other new programs, have funds dedicated to innovative projects that address these themes. There is a myriad of funding opportunities available if the astute researcher follows trends and themes at NSF. Investigators also need to be flexible in how they envision their project and special emphases (e.g., working with teachers) that may not previously have been part of their research agendas. In contrast, however, those researchers who stick to their comfort zone and continue to do business as usual will likely be left behind.

Integration

The concept of integration of research activities is fundamental to the advancement of science in the twenty-first century (Honey et al., 2014). Since Bush's (1945) report, scientists have understood that fresh ideas and innovations are promoted by a diversity of perspectives and approaches. Thus projects that are termed integrated, interdisciplinary, or multidisciplinary, each with subtly different meanings (English, 2016), oftentimes add value to the research enterprise. The National Academy of Sciences (2014) issued a report on a workshop that brought together academics from a variety of disciplines with the purpose of furthering STEM and integrating art and design into the mix – which has been termed STEAM. This is just one example of initiatives that realize the importance of integration as a generator of innovation.

STEM integration is of prime importance in K–12 education, particularly with the development of the new Next Generation Science Standards (NGSS, 2013). The so-called 3D (multidimensional) learning includes disciplinary core ideas, science and engineering practices, and cross-cutting concepts. For example, within the cross-cutting concepts, scale and proportion are fundamental to many STEM content domains. Krajcik and Delen (2017) discuss the importance of learning STEM in K–12 through the connection of science and engineering concepts and practices. They also note that students who engage in STEM will have the opportunity to develop twenty-first-century skills, including problem solving, communication, and collaboration (Krajcik & Delen, 2017: 21).

Historically within the K–12 educational framework in the United States, science and math have largely been taught in isolation from one another. In the twenty-first century, however, there has been a push to integrate STEM education in the United States, so that innovative programs that combine, for example, math and science, or reading, writing and science are viewed as being important to engage learners at all levels of K–12 and even in higher education. In 2015 I visited a fourth-grade class in Santa Cruz, California, during their reading time when their lesson was about dinosaurs. The students seemed excited, and any boredom of doing routine reading in isolation was not evident in that classroom. As we will also learn in Chapter 11, integration is a fundamental part of NGSS (2013) within the domain of "cross-cutting concepts."

Case Example: 3D-Printed Fossils and STEAM

With regard to STEM integration, the iDigFossils project (2017) funded by NSF's ITEST (Innovative Technology Experiences for Students and Teachers; NSF, 2017g) program engages scientists, teachers, and students in using 3D-scanned and -printed fossils to develop innovative lessons that integrate STEM. The **S**cience is paleontology and evolution through time, the **T**echnology and **E**ngineering are integrated through the use of 3D scanners and printers, and the **M**ath is incorporated via intentional lesson planning. Grant et al. (2017) explain this program by using the giant fossil shark, *Carcharocles megalodon* or "Megalodon" (Box 3.1), but other

Box 3.1 3D Megalodon, Integration, and STEAM

1. Collecting a 10 million-year-old giant shark (*Carcharocles megalodon*) tooth from Panama (Scott Flamand photo).
2. A student studying a 3D-printed *C. megalodon* tooth as part of a lesson plan.
3. An example of STEAM. Middle-school students in Tampa, Florida, worked with both their science and art teachers to reconstruct a life-size jaw of *C. megalodon*.
4. Scanning a *C. megalodon* tooth using a laser desktop scanner (FLMNH Jeff Gage photo).
5. 3D-printed *C. megalodon* teeth. (Figs. 2,3,5 are courtesy of M. Hendrickson and Academy of the Holy Names, Tampa, Florida.)

lesson plans resulting from this project teach about giant snakes, fossil horses, and human evolution (iDigFossils, 2017). At several schools, art has been added so that STEM becomes STEAM through the integration of project-based lesson plans in both science and art classes that yield life-size reconstructions of fossils like Megalodon.

Prior to Grant et al. (2017), little had been done with the application of 3D scanning and printing of fossils for classroom use (Hasiuk, 2014). It cannot be disputed that the 3D wave is upon us and has permeated society. This can be seen not just in teaching and learning, but in many other industrial and medical applications. Likewise, the concept of maker spaces has resulted in many libraries renovating spaces for 3D-printing and other hands-on activities. A decade ago, few high schools and libraries had access to 3D-printers, and although by no means ubiquitous in the United States today, they are becoming more commonplace. Although Grant et al. (2017) and the iDigFossils project primarily focused on fossils in middle and high school, Cox et al. (2019) are using 3D fossils to develop lesson plans about fossils in elementary schools. The 3D movement is revolutionizing access to physical specimens such as fossils for classroom instruction. Now a teacher can do "one-stop-shopping" through cloud-based portals that allow not only the downloading and printing of lesson plans, but also the scanned image files to be 3D-printed. Although we will revisit K–12 Broader Impacts in Chapter 11, the case example presented here (Box 3.1) illustrates the power of STEM integration, also including Art (STEAM).

Summary and Concluding Comments

All of the three concepts described in this chapter – innovation, opportunity, and integration – are fundamentally important both to developing a more successful proposal strategy, and also to potentially increasing societal benefit. However, many proposers do not understand the importance of these criteria in setting their project apart from the many other competitive proposals submitted to NSF. Since the middle of the twentieth century and the beginnings of NSF, it has been realized that innovation is a cornerstone of scientific advancement. However, at the same time, innovation does not always come easily and is difficult to quantify. How can one say, for example, that one project is more innovative than another? Scholars have come together to address this matter, but so far it requires output and metrics that are oftentimes developed years after the end of a project funding period. It takes thought and brainstorming to specifically address new ideas and methods; this is not something that is explicitly described or understood during the proposal development. With regard to opportunity, many scientists continue to go back to the same NSF programs within their comfort zones. This rut is despite the fact that many of these programs, with success rates below 10 percent, usually require innovation or a special "hook" to be funded. The nimble researcher, however, will look for special opportunities at NSF as new programs are developed and implemented. NSF issues hundreds of Research Fellowship Program-style solicitations and Dear Colleague Letters each year. These oftentimes represent those special opportunities not necessarily anticipated or that cannot be planned for over the long term. Finally, integration, particularly if it includes previously disparate parts of STEM, is also likely to produce more interesting projects worthy of funding. STEM integration, and even STEAM, are trendy topics within educational circles today (McGinnis, 2017), and the application of this concept is also important to basic research in the twenty-first century, particularly with regard to NSF's Broader Impacts.

4 Communication and Dissemination

The Elevator Speech

An "elevator speech" is a short, intense, and mostly one-way conversation intended to convey essential information. In a sense, it is a verbal abstract composed of sound bites lasting about a minute or two. The elevator speech is an important communication and networking tool.

In the fall of 2009 I was a program officer at NSF. One day on my way to work, I entered the lobby elevator. As the doors closed, I realized that the only other person in the elevator was a distinguished older gentlemen dressed in a business suit. I recognized him as Arden Bement, the NSF director. He looked at my NSF badge and, after he introduced himself, he said: "We have never met, what do you do here?" Knowing that I had about eight floors (to where I worked) to finish the communication, I responded: "I am a program officer on leave from the University of Florida and work in EHR [Education and Human Resources]. I manage an informal science education portfolio focused on how children and families learn in museums, science centers, and other similar institutions. I am also involved with another program that funds researchers wanting to communicate to public audiences." The elevator stopped at my floor (he was likely going up to the thirteenth floor) and as I got out, Dr. Bement ended the conversation with "keep up the good work and have a nice day."

> Communicating science effectively, however, is a complex task and an acquired skill.
>
> NAS (2017b)

Introduction

During my Broader Impacts graduate seminar, students are assigned an exercise in which they video themselves doing elevator speeches. We then play these back and the entire class critiques what they have said. Listening to the groans beforehand, it is clear that most of the students do not enjoy this experience. (Although, in the course evaluations they later reflect that it was valuable for them to do the elevator speech.) In today's world of sound bites and a few hundred characters (Twitter), the importance of focusing what you say into short dialog cannot be overemphasized. This style of communication is not restricted to your colleagues. It happens all the time, for example, at social events, in chance conversations in airports, to administrators and politicians working a crowd, to potential donors, and even a short conversation trying to "pitch" a research idea to an NSF program officer. The National Science Foundation (2017d) specifically states an expectation for an elevator speech statement in one of the grant programs. If you cannot encapsulate what you say into an elevator speech, you are missing a basic communication tool.

Sometimes a distinction is made between the terms "communication" and "dissemination." According to Leitat (2018), communication is focused on the promotion of projects and their results to a broader audience, such as the general public, using non-scientific language. Examples of communication channels include traditional mass and broadcast media and, more recently, social media and websites. In contrast, dissemination is the public disclosure and presentation of project results, particularly to professional peers, stakeholders, and policy-makers. Dissemination is typically done using scientific language via peer-reviewed journals and professional conferences. These distinctions are sometimes blurred – for example, the dissemination requirement for NSF Broader Impacts plans can also include communication to public audiences. In this book, the distinction between these two terms is somewhat arbitrary and fluid.

As the quote above indicates, there is no question that effective science communication is a skill that is not easily acquired "on the side." For some scientists, communication comes naturally, whereas for others it remains a challenge. The recent report published by the National Academies of Sciences, Engineering, and Medicine (NAS, 2017b) underscores the importance of effective science communication in modern society. As described below, many ways exist in which we can disseminate knowledge and communicate more broadly.

Academic Discourse, the "50-Minute Hour," and Related Anachronisms

Many college lectures today are deemed dull – and with good reason.

Gross-Loh (2016)

More than a half-century ago, Lindner (1955) wrote the book *The Fifty-Minute Hour*, which has now become a metaphor in psychoanalysis for the amount of time in which face-to-face counseling is typically done with patients. So too has the 50-minute hour become synonymous with learning intervals in academia, from the standard class-period lecture time in higher education, to the weekly seminars and colloquia of academic departments on college campuses. The lecture format, which has been around for centuries, can trace its roots back to the Greek oratory style (Gross-Loh, 2016). A common theme of these is typically one-way communication in which the professor (expert) communicates knowledge to students or the interested audience (novice). The problem with this model is that learning research shows that the 50-minute lecture format is an anachronism, a type of pedagogical communication from times past. Unless they are used with inspiration (Gunderman, 2013), or with some form of active learning (Vise, 2012), lecture formats simply do not enhance engagement and learning. This 50-minute hour format also has been pervasive in K–12 education, although some schools have transitioned to longer blocks of class time that allow for more in-depth, and oftentimes inquiry-based or self-directed, learning. I have transitioned to two-hour blocks of time in most of my graduate seminars. This format is simply less rushed, particularly when a presentation is followed by discussion.

The academic seminar is another means of communication that has passed its effective lifetime in many circumstances. With time being a limited resource, one could spend much of the week on college campuses attending academic seminars. I go to very few seminars, mostly because I do not want to invest the hour to learn what is being delivered. Many seminars tend to drone on, present too much data and complex graphs that are difficult to absorb before the next slide, and the punchline or take-home message is at the end. I typically only attend seminars when the title sounds interesting, the topic is something close to my research, or to support a colleague or student. This stodgy point of view is perhaps missing an opportunity, but there simply is not enough time to do everything.

Professional scientific meetings provide another opportunity for learning via cognitive exhaustion. A typical format is a 12-minute talk, followed by 3 minutes of questions (if the speaker does not run over time), and this format continues, oftentimes unabated, for 3 or 4 days. By the middle of the first morning of this format many of us are looking for the door, coffee table, exhibitors' hall, or some colleague to network with rather than return to the unidirectional learning opportunity. Temporarily re-invigorated with caffeine, I then select talks strategically. This lack of dedication stems from the idea that if the topic is really interesting, I can read the published version. Participants who attend every talk during the meetings, and come away refreshed and invigorated with new knowledge, are to be commended.

I am likely not alone in my aversion to long, methodically developed talks in classes, seminars that drone on, and meetings with mind-numbing short talks with no break in format. All of these have a theme in common: unless they are particularly imaginative, these styles of discourse are not well suited to modern society. Professional communicators, e.g., in science journalism, have realized that the typical way that scientists communicate is not what most people want (AAAS,

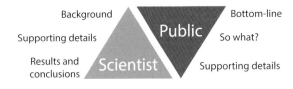

Fig. 4.1 The different verbal dissemination and communication styles of the scientist (left) versus a public presentation (right, modified from AAAS, 2010).

2010). Instead of a formal talk organized via the methodical development of background, supporting details, methods, data analysis, and results/conclusions (Fig. 4.1), the talk could be flipped to make it more interesting and likely engage a broader audience. Thus, the talk would include the "bottom-line" (the key idea or punchline), "so what?" and supporting details. The elevator speech concept is, in a sense, the bottom-line and "so what?" components of this condensed form of communication. If the listener wants more (supporting details), then other ways to acquire this information can be used, such as follow-up via email or access to a publication.

Jargon and the Scientific Meaning of Words

Another matter that should be of prime concern to scientists is the audience that they intend to reach. Although we will discuss intended audiences more in Chapter 8, here we explore these in the context of the pervasive use of jargon. One of the easiest ways to turn off your audience is to communicate a smorgasbord of jargon; this will instantly form a learning barrier. Sometimes, scientists say that they hate to "dumb down" the topic for a general audience, but this is a particularly myopic and egocentric view of the world. In contrast, it should be realized that most walks of life have their distinct vocabulary. If the intent of the scientist is to communicate effectively, they must predict their audience before a presentation. It is easy to understand that an audience of graduate students in one's own discipline will have a different set of prior knowledge than attendees at a general public lecture. On the other hand, there are subtler nuances that should be realized. A seminar presented in a biology department ought to have a different emphasis from one with similar content presented to a geology department (Box 4.1).

Words in general use by scientists on the one hand and the public on the other hand potentially have different meanings to these two audiences. Take, for example, the word "theory." To the scientist, evolution is a theory, or an explanation for a natural phenomenon, that is backed up by evidence. Science changes, as do theories, and scientists understand this. The problem with the term "theory" with the general public is that they can dismiss evolution by saying, "Oh it's just a theory," or it is a hunch, or speculation (Hansen, 2012), implying that it is not well substantiated. Another example is confidence intervals and probability. In courts of law, when quantitative forensic evidence is presented with probability limits (e.g., 95 percent confidence), the opposing counsel might say: "Well, they are not really sure about this conclusion." Again, this kind of statement comes from a misunderstanding about how probability and statistics work and the concept of (un)certainty in science.

Box 4.1 Knowing Your Audience: The Nuances of Words, Emphasis, and Jargon

With the advent of computer-generated visual "slide" presentations (e.g., PowerPoint), many of us spend hours developing talks that are oftentimes then recycled for another presentation. Unless the audience is essentially the same, it is not advisable to present the exact same talk. Some aspects of talks become stale and can be improved with at least minor modifications. After my presentations I sometimes make annotations about what seemed to work, what did not, and how I would change the slides next time. Two examples are presented here about nuances of presentations related to presumptions about prior knowledge and interest of the intended audience.

  Different audiences, similar talk. The core of the two talks presented here had many of the same slides related to our teacher professional development project in Panama. Some slides were used on both talks verbatim; others had minor changes (e.g., amount of jargon). **First.** This title slide was used for a talk presented to a Rotary Club in Santa Cruz, California. This club had civic leaders and members of the local school board. I judged that the audience might be more interested in the local impacts accrued back to the schools and their students. **Second.** This title slide was used to present a seminar at my home institution, which is a diverse museum audience representing different disciplinary collections, faculty, students, and staff. In the latter talk, I perceived that it was more important to discuss the research focus (Great American Biotic Interchange) and the Research Experiences for Teachers (GABI RET, 2017) program in which participants went to Panama to collect fossils. The core of each of these two talks had a series of the same slides.

Fossils in Santa Cruz K-12

Bruce J. MacFadden, Ph.D.

Visiting Scientist
Santa Cruz County
Office of Education
Santa Cruz CA 95060

Distinguished Professor
Florida Museum of Natural History
University of Florida
Gainesville FL 32611

Box 4.1 (cont.)

GABI RET
Great American Biotic Interchange
Research Experience for Teachers

Bruce J. MacFadden

Florida Museum
of Natural History
University of Florida
Gainesville FL 32611

Same slide, different audiences. I have given a seminar titled "Fossil Horses: Evidence for Evolution and Exhibits" to several different academic audiences. One of the introductory slides (shown here) depicts the 10 kinds of evidence for evolution. To a biology audience, all of these, with perhaps the exception of two fossils of the early bird *Archaeopteryx* (lower row, second from right) and the fossil horse family (lower row, first on right), are easily recognizable. When giving this talk to an earth science audience, I spend a little more time explaining the other eight examples.

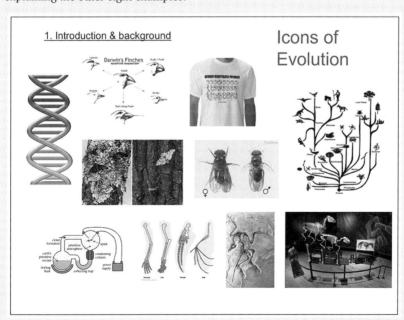

It is therefore important that scientists who want to effectively communicate are sure to: (1) understand their audience; (2) minimize jargon, and if it is critical to the presentation, stop and explain the word(s); and (3) for public talks, understand that the meaning of some words can be different in a strictly scientific context versus the understanding of a general lay audience.

Scientific Writing and Dissemination

Just as verbal presentations have undergone change, so too has science writing in some circles. Some journals no longer expect the fundamental component of the scientific method taught in schools – the Materials and Methods section – to be integral to the flow of text. For example, the journals *Science* and *Nature* relegate these, and oftentimes the primary data that back up the results and interpretations, to Supplementary Documents. This is actually not a bad idea because most readers care less about the materials and original, lengthy, data tables. As such, valuable real estate is conserved in the main article itself. A professor in graduate school once told me (paraphrased):

> I almost never read a scientific article from start to finish. I look at the title to see if I am interested, who wrote it, and then I read the abstract. If a more in-depth analysis is in order, I look at the section headings, figures and tables (but not if the latter has a large data matrix), and speed read or scan the text. A quick glance is oftentimes in order, and then I'm done.

The novice scientist can benefit from this advice, as I have since my graduate student days. Some journals, like *Trends in Ecology & Evolution*, separate important supporting text, like definitions or small thematic units, to boxes in the margins. Other journals use graphic means, such as changes in fonts and colors, to help the reader understand what is important.

Some high-profile journals with broad readership like *Science* have gone to different sized contributions. These range from a few-sentence synopsis ("In Brief") of research published in *Science* and other journals, to longer high-profile "Research Articles" of a more synthetic nature, followed by shorter "Reports" of mostly single-study results. Occasionally a "Research Article Summary," which is typically limited to one page, is linked to a more complete article online. In the earth sciences, *Geology*, which limits all research articles to four printed pages, has been a popular and successful format. This journal has a relatively high impact factor (also see Chapter 5) of 4.5 within its discipline and for the past decade has been the #1 professional journal published in the field of geology (GeoScienceWorld, 2017).

Poster Presentations

If done properly, posters can be an effective way to communicate science. Thus, in addition to the standard 15-minute talk, science is frequently communicated via posters at professional meetings. Posters are "widely used in the academic community. They summarize research in a concise and attractive way to communicate essential information to publicize your work and hopefully generate discussion" (NYU, 2017). In many respects, this method of communication at big meetings can actually be preferable in terms of a meaningful dialog between the presenter and audience.

On the first day of a big meeting, after the few talks that I can withstand, I typically leave the session of platform talks and symposia to find my way to the exhibition hall, which is oftentimes the venue for poster presentations later in the day. In some larger meetings, hundreds of posters are already on display, usually organized by thematic units. In the morning before they are officially scheduled to be presented it is quiet and there are few other people in the hall (they are mostly elsewhere listening to talks). This is a wonderful time to do a quick run through of the posters, read those that seem interesting, and make plans to return when the author is scheduled to be in front of it. I usually tell the poster presenter that I previously came by and looked at the poster and would like a brief synopsis from them. This can then move into additional conversation. One of the most memorable meetings that I have ever attended was on fossil bone chemistry in Cape Town, South Africa, in 2005. The poster session was held during a "happy hour," and all of the participants went around en masse from one poster to the next. With drinks in hand, we listened to elevator speeches from each presenter and then asked questions. It was a wonderful form of two-way dialog. This method of presentation is not practical for very large meetings, but it certainly was fun for the ~ 75 scientists that attended the Cape Town meeting.

In addition to how the poster is presented, the graphical organization is also important to its successful engagement by the intended audience. Innumerable resources and best practices (Inset 4.1) are available for how the scientific poster should be developed (e.g., CCMR, 2017; Purington, 2017). As scientists, we also can learn much from the graphic design and marketing view of the world (e.g., Cousins, 2015). We will talk more about museums and exhibits in Chapter 13, but a brief mention of how interpretive panels are designed is relevant here. Whether it be a poster or a museum exhibit, the common thread and intent is to capture the engagement of the audience. Learning research (e.g., Serrell, 1998; Yalowitz & Bronnenkant, 2009) about "time-tracking studies" has shown that the great majority of museum visitors spend less than a minute at a particular exhibit module, such as a graphic panel or small display. The designer

Tips for poster design

- Scale to be viewed from 3 meters (will mostly be viewed, not read)
- Large fonts, limit to a few kinds, do not use all caps fonts, most fonts greater than 24 point
- Use accessible colors
- One-sentence abstract under title
- Big question, discoveries, and results up front rather than near the end
- Text better as bullet points, rather than standard paragraphs
- References tiny (small font) at the end
- Logos more important than written acknowledgments
- Add social media contacts, websites, and other links

Inset 4.1

Fig. 4.2 Butterfly Rainforest at the Florida Museum of Natural History. Museum interpretive panel showing how information is presented to engage different kinds of audience styles. This has many of the features that are also characteristic of an effective poster (Inset 4.1). For example, the title and one-sentence abstract convey the overall content and "big picture" of the panel. From there, most of the visitors move on ("streakers"), whereas this panel also provides more content for those who want to learn more ("strollers"), or as much as the panel has to offer ("studiers"). Other strategies employed in this panel include different size fonts and colors, asking questions rather than stating facts, a mix of graphic images and text, and the use of "bullet points," in this case with the three butterfly symbols (FLMNH, Stacey Breheny graphic).

and content curator of the exhibit therefore has a limited amount of time to communicate the most essential information. Scientists who do not want to "dumb down" their content and fill the graphic panels with lots of information are missing the boat. Like it or not, a large percentage of their audience will simply not read long passages of text. At the Florida Museum of Natural History, we have developed a tripartite strategy based on what we know about visitor behavior and engagement with the exhibit panels. We refer to this as the "streakers, strollers, and studiers," with each of these kinds spending an increasingly greater amount of time in front of the panel (Fig. 4.2).

Verbal, Visual, Written, and Hands-on Science

A tremendous body of research and literature has been devoted to better understanding how people learn; in fact, the field of educational psychology studies this topic. It would be impossible to cite all of the relevant publications on this topic; however, Bransford et al. (2000) seem to provide a comprehensive and authoritative place to start. People learn differently at different stages of their development. Novice and expert learners learn differently. People learn with different senses, and have different "learning styles" (although colleagues of mine in the College of Education bristle at the use of this term because it is too simplistic). Nevertheless, Chick (2017) defines learning style as "how learners gather, sift through, interpret, organize, come to conclusions about, and 'store' information for further use." She uses the acronym VARK for these styles, which encompasses **v**isual, **a**ural, ve**r**bal, and **k**inesthetic learning. She also notes that this is only one scheme and there are more than 70 other different learning schemes. Although this would seem counterintuitive, the disappointing thing about learning styles or schemes is that it is unclear the extent to which matching learning styles to learners is effective (Chick, 2017). Nevertheless, the assumption is that people learn differently for a variety of reasons. As educators we typically try to use a variety of strategies, if not to improve learning, perhaps to increase engagement. Thus, inquiry-based (learner focused) and hands-on (kinesthetic) are popular learning strategies. These strategies have universally increased in both formal classroom as well as informal learning environments. The following represent a few examples of how we can do a better job of communicating and possibly enhancing the learning experience.

Infographics

Infographics typically combine graphs or graphic images, sometimes photos, and text to communicate information or make a point in a way that will be more engaging than any of these means separately (Fig. 4.3). The marketing and public relations industry has understood this strategy for a long time. As a consequence, they are further ahead of scientists in the way they engage their audience, and we can learn from them. Professional organizations such as AAAS understand this, and many illustrations in *Science* are infographics in their design. Nevertheless, the use of infographics has, despite their effectiveness, been slow to permeate other formal scientific literature. We mostly cling to the tradition of separate graphs, photos, and

Value of an infographic

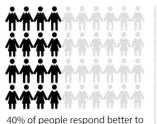

40% of people respond better to visual information than text.

High-quality infographics are 30 times more likely to be read than text articles.

Fig. 4.3 An infographic about infographics (adapted from Customer Magnetism, 2017).

text to illustrate peer-reviewed publications, when in fact a place could be made for the infographic in this kind of dissemination of science.

Similarly, presentation graphics such as those created with PowerPoint also lend themselves to, and effectively promote, the use of infographics. The concept of a "graphical abstract" now used by some science organizations is another example of an infographic. Weidler-Lewis et al. (2018) also describe the value of infographics in promoting creativity in K–12 classrooms. In summary, although infographics are primarily associated with the commercial sectors of the public, these can be of much value for science communication and learning in different venues and to diverse audiences.

"PowerPoints"

Just like to "google" something has become a verb, a "powerpoint" is widely used as a noun, although it is one of several kinds of presentation software programs. In the days before PowerPoints, we used a series of 2 × 2 inch transparent slides (and hence the name of each PowerPoint image), or clear acetate overhead sheets with content on them. The slide and overhead projectors, mainstays of previous generations, have mostly gone the way of the typewriter and slide rule. PowerPoints have thus carried us into the computer age. If done properly, they can enhance the quality of the presentation and the organization of your content. If done poorly, they can provide a sedative to induce a nap during a seminar. PowerPoints also seemingly preclude the need to take notes if the presenter lets the audience know that the talk will be made available to them. However, studies have shown that the process of writing notes can be a metacognitive means of learning (Boch & Piolat, 2005). Although my presentations are frequently made available, in classes for example, I still encourage my students to take notes to promote learning.

Despite their universality, PowerPoints can be done well, or badly. In one view, they can function like individual infographics, combining images (photos and graphs, for example), text, animations, and even embedded videos. Just like for posters, scientists are typically not taught how to do PowerPoints. Many resources exist about how to design and produce effective PowerPoints (e.g., Krogue, 2013; NCSL, 2017; Wax, 2017). Certain practices are simply not effective. Perhaps the most egregious of these is a detailed series of slides loaded with text that the presenter reads – the kiss of death for engaging your audience. Likewise, a slide or slides crammed with so much content that the viewer has instant cognitive overload is also a bad practice. Another bad idea is a presentation in which too many slides are crammed into the talk – for example, 50 slides for a 12-minute

PowerPoint presentation tips

- Minimize text
- No more than 6 bullet points per slide
- Complete sentences not necessary
- Either sans serif or simple serif fonts
- Consistent slide layout and design
- Slide background simple or muted
- Animations can be distracting
- Minimize complex data slides and graphs
- Talk to the audience, not the screen
- Do not read text like reading a book
- About one slide per minute

Inset 4.2

talk. Certain fonts are considered better than others. In addition to, or along with, these points, years of presentations, both before and after the PowerPoint era, have helped me to develop a series of best practices and tips (Inset 4.2). These are based not on any formal training in design or presentation, but my "on-the-job" experiences about what seems to work. In addition to these tips, a colleague from the journalism world has told me that, in his field, the trend is to minimize, or eliminate, text from PowerPoint slides and let the talk flow with graphics and images.

Dissemination: "Publish or Perish" Is Not Enough

Broaden dissemination to enhance scientific and technological understanding, for example, by presenting results of research and education projects in formats useful to students, scientists and engineers, members of Congress, teachers, and the general public.

NSF, Dear Colleague Letter (2007)

The pursuit of science is not complete without dissemination, which is also considered essential for a comprehensive Broader Impacts plan. It will come as no surprise that the gold standard of the dissemination of research results is via peer-reviewed articles in professional journals (the maligned concept of "publish or perish"), followed by presentations at professional meetings. With regard to NSF's Broader Impacts merit review criterion, even this is just the beginning. The dissemination strategy described above relates to the professional science community, but other potential audiences could be included, as makes sense to one's funded research projects. For example, the quote above mentions Congress, teachers, and the public. Of these, Congress is important because they pay the bills and for the possibility that scientists might positively influence governmental policy and views of science. K–12 teachers are important because of their potential impact on the next generation of students. The public is important because of the overall impact and societal benefit that can result from effective communication with them. It should also be realized that these audiences are neither homogeneous nor monolithic. Within these audiences we should be mindful of the impact that can be had for underserved minorities through the dissemination of our scientific discoveries.

The various ways in which effective dissemination can be achieved are likewise manifold, and can include press releases, social media outlets (e.g., Facebook and Twitter), websites and blogs, public talks (e.g., Rotary and science cafés), public participation (e.g., community science), and museum exhibits. This list is not meant to be exhaustive, and likewise it is not intended for the researcher to implement all of these dissemination modes. It is best to choose several of these and do them well with appropriate audiences. Thus, via widespread and effective communication, as scientists we can support the notion of social responsibility.

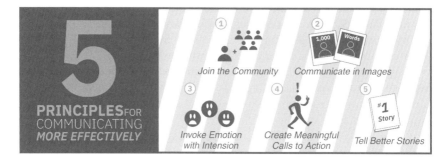

Fig. 4.4 Five tips for effectively communicating, including storytelling (Kays, 2018).

Science Communication as Storytelling

> There is a narrative that scientists are not good communicators. That they're content to tuck their work away in journals and share it at sparsely attended academic conferences … I don't believe that narrative.
>
> Christiano (in Kays, 2018)

We recently interviewed candidates for a communications manager position in our museum. Several referred to themselves as science storytellers. I had not thought of science communication in this way before, but it obviously makes sense. Many scientists are comfortable presenting their results using the formal scientific method, but this form of communication can be boring to the public, and other modes may be more effective (Fig. 4.1). Although storytelling is not usually associated with science, Dahlstrom (2014) advocates for science communication to be considered within the framework of story-telling. As such, it becomes a narrative that can potentially be more engaging outside of the ivory tower.

Coincidentally, while I was completing this chapter, an article appeared in *Explore*, a magazine promoting research done at the University of Florida. One article, titled "Communicating science for good," features science communication professionals, including Professor Ann Christiano, and programs in our College of Journalism and Communications (Kays, 2018). In that article, five tips for effective science communication are presented, including telling stories (Fig. 4.4).

Wrap-up: Communication and Broader Impacts

> Scientists spend most of their working life writing, yet our writing style obstructs its key purpose: communication.
>
> Doubleday & Connell (2017)

This chapter was intended to be an "on-the-job" view of science communication. With a general topic this broad, it is neither comprehensive nor deep. I am not a science communications expert, so the topics covered above are focused on my own interests and experiences. In the twenty-first century, other important means of communication such as social media (e.g., Facebook and Twitter) and YouTube are widespread communication strategies that will be covered in Chapter 15. Science communication is thus a theme that weaves together many topics covered in this book. The more effective we are at both dissemination to our peers, and more broadly communication to the general public, the more successful we will be as scientists.

Missed Opportunity: Communicating Research to Public Audiences

The National Science Foundation expects investigators to be effective science communicators; this is particularly the case with a successful Broader Impacts strategy. The problem, however, is that few scientists have any training in communication. For about two decades a program existed at NSF in which principal investigators who took a particular interest in Broader Impacts and outreach could apply for separate funds that would promote science communication and outreach. This program started in the mid-1990s as informal education supplements to existing research projects, and ultimately morphed into a separate program titled: "Communicating Research to Public Audiences" (CRPA; NSF, 2003). As an NSF program officer in 2009–2010, I was in charge of this program, which by then awarded up to $150,000 to worthy projects. My time at NSF taught me that CRPAs were the best-kept secret within NSF's research directorates; they likewise were largely unknown to the rank-and-file scientists at their home institutions. They had many positive outcomes in terms of communicating innovative research. An unintended outcome for the recipients of these awards was the opportunity to hone their science communication skills. It thus greatly benefitted not only the target audience of the project, but also provided "on-the-job" communication experience for the scientist. Unfortunately, despite the innovation of many of the projects (Korf, 2015; Fig. 4.5), the CRPA program was never very popular in terms of the number of proposals submitted. It ultimately ended and was not replaced by another program that might have encouraged professional development of communication and related strategies for scientists.

Final Comments

A few take-away points of this chapter are as follows:
1. Effective communication underlies every aspect of Broader Impacts. The more you understand about how science can be communicated, the more successful you will be in optimizing these efforts.
2. This chapter opens with an anecdote about elevator speeches. The importance of brevity in communication cannot be overemphasized. A general rule of thumb is the briefer, the better. In so doing, you will appeal to a larger audience as you deliver the "punch" or essence of the intended communication.

Fig. 4.5 AstroDance. In this NSF CRPA project, the fundamental laws of astrophysics are taught to deaf learners via dance. In this photo, dancers use stretch bands to represent gravitational waves from supernova; Pathmanathan, 2014; also see AstroDance, 2018 (Erin Auble and the AstroDance team photo).

3. One does not have to aspire to be a science communications professional, or have formal training in this field. Nevertheless, a better appreciation of communication best practices will enhance your science and increase the probability of successfully getting the message out, as well as developing a better Broader Impacts plan for proposals.

5 Promoting Yourself and Optimizing Impact

Charismatic Fossils and Their Impact

I study fossil horses. This is fortunate because people love horses, so engaging my audience is easy; they already are interested in what I have to say. If I studied foraminifera or titanotheres, for example, I would need to spend time convincing my audience that these groups are worthy of their attention. Charismatic fossils have a public appeal, making them instantly interesting. No kind of fossil is more charismatic to kids than dinosaurs. Also vying for top billing is the giant, extinct predatory shark "Megalodon" (*Carcharocles megalodon*). Just like dinosaurs, human fossils, and horses, Megalodon has been the object of much recent scientific research. This charismatic species has also engendered much hype in the popular media; for example, the speculation that, like the fabled Loch Ness Monster (thought to be a plesiosaur, a holdover from the Age of Dinosaurs), Megalodon is still alive today.

Several of my graduate students have studied fossil sharks, including Megalodon. In particular, Catalina Pimiento investigated aspects of the biology and extinction of this species that had enormous impact and reach. In 2010 she led a study looking at evidence for a shallow-water Megalodon "nursery" in Panama during the Miocene, 10 million years ago (Pimiento et al., 2010). She later published a statistical model showing that despite claims to the contrary in the popular media, Megalodon is not living today and became extinct during the Pliocene, about 2.6 million years ago (Pimiento & Clements, 2014). Taken together, these two articles on Megalodon have been viewed online more than 100,000 times. A follow-up article about ancient distributional patterns of Megalodon (Pimiento et al., 2016) has an astoundingly high Altmetric score (defined below) greater than 350. This is in the 99th percentile for all articles cited in the *Journal of Biogeography*. People are generally fascinated with body size, and how this feature evolved, particularly when this process resulted in large species. A colleague of mine called this the P. T. Barnum effect, in reference to the nineteenth-century entrepreneur who would display large exemplars, some of which were hoaxes, in his traveling circus. In a scientific context, the concept of "Cope's Rule" is a pattern in which body size tends to increase from ancestor to descendent species (e.g., Stanley, 1973). With estimated body lengths of about 15 meters, Megalodon was the largest predatory shark that ever lived. In collaboration with another graduate student, Catalina published a paper in the high-profile journal *Paleobiology* that provided a model for how the Megalodon's body size evolved through time (Pimiento & Balk, 2015). As of late 2018, this paper has a very high (for the journal) Altmetric score of 134. Catalina did not start her PhD fully understanding Megalodon's appeal, but as she continued her studies it became clear that this charismatic fossil species is a winner.

Scientists select their research for a variety of reasons. Although it should not be an overriding factor, it does not hurt to work on a project, problem, or creature that is in the limelight. It seems that this impact extends not just to the public, but also has a feedback loop to other scientists as well, who are keen to jump on the popularity bandwagon.

Introduction

> vertebrate paleontologists are to be compared to the conies of scripture—"a feeble folk, that dwelleth among the rocks."
>
> Romer (1969)

Although perhaps not as humble as the famous twentieth-century Harvard paleontologist Al Romer, many scientists are content to make their research discoveries and disseminate these to their peers with little additional fanfare. We just are typically not the kinds of personalities who would strategically promote their careers to broader audiences. To be sure, there are exceptions to this rule. In the previous generation, Carl Sagan and Steve Gould were champions of science, as Neil deGrasse Tyson and Bill Nye "the science guy" are today. It is becoming increasingly clear that with the competitive environment in the research world, scientists need to promote what they do, not just to their peers, but also to funders and the general public. Clear strategies and obvious best practices are well known to many scientists; however, in the rapidly evolving world of cyberenabled technology, other strategies are emerging as well. Any astute scientist in the twenty-first century is well served to understand the value of optimizing their impact.

Promoting Yourself to Your Professional Peers

Peer-Reviewed Publishing: How to Choose a Journal

A universal aspiration of most scientists is to "get published" in the highest-cited peer-reviewed journal possible. Decades ago, when I started publishing (my first peer-reviewed paper was published in 1976), we understood that publishing a paper in *Science* or *Nature* was something that we should aspire to do. This would result in widespread dissemination within the broader professional community; however, we also realized that this was a challenge. If one was unsuccessful in either of these journals, which happened frequently, then a solid back-up plan was to publish in one of the most prominent journals in one's respective fields – in my case, for example, the *Journal of Paleontology*. None of this process was quantified and no metrics existed to guide decisions. It was simply understood as a best practice and part of our professional culture.

Regardless of where our papers are ultimately published, we only have so many articles that can be written in one's academic career. Typically, a total of 50–75 peer-reviewed papers is a general indication of solid output from a productive scholar. This also equates to two to three papers per year. Therefore, only a finite opportunity exists to make one's professional impact. One way of optimizing this impact is by choosing top-cited journals. Likewise, methods for selecting these journals nowadays is far more of a deliberate process, one based on impact metrics. As such, the landscape for optimizing impact in the professional arena has changed. Now we know that journal impact factors validate what we suspected before. Journal impact factors (JIF) are calculated as the yearly average number of times that articles published in a particular journal are cited. Thus, with 2016 impact factors of 40.137 for *Nature* and 37.205 for *Science,* these two journals lead the pack within the wide-ranging field of general science. These are followed after quite a gap by 9.661 for *Proceedings of the National Academy of Sciences.* Typically, after unsuccessful attempts to publish in these journals, we then tend to go to the leading or flagship journals within our respective fields, which in paleontology include *Paleobiology* (2.886) and the *Journal of Vertebrate Paleontology* (2.346; Scijournal, 2017).

Comparing impact factors across disparate fields is like comparing apples and oranges; it simply does not make sense. Few other disciplines have the impact of the biomedical sector, which dominates the highest rankings, with, for example, the well-known *New England Journal of Medicine* sitting at the top at 72.406. The majority of journals with an impact factor greater than 20 are from the biomedical sector. In specialized fields of basic science, like paleontology, none of the journals have impact factors much above 3. Regardless of discipline, in some circles publishing in low-impact journals is given less weight in academic evaluations. Alberts (2013a: 787) notes that in some fields publishing "in a journal with an impact factor below 5.0 is of zero value." This practice is unfair to disciplines with a smaller number of practitioners, such as paleontology, in which even the most highly respected journals have relatively low JIFs. I once attended a workshop at which the organizers had contacted a journal to publish the proceedings as an issue of peer-reviewed contributions. Although well respected in our field, the JIF was just under 2.0. Some of our European colleagues lamented that, given the low ranking, they could not publish in this journal because the article would not count fully toward promotion.

Like many attempts to quantify academic productivity, although initially based on logical reasoning, certain practices have been abused or gone to extremes that make little sense, or are simply unethical. With regard to JIFs, in 2013 the DORA (Declaration on Research Assessment) had sweeping recommendations, including the ill-advised use of JIFs to judge individual scientists' work (Alberts, 2013a; Hoppeler, 2014). Thus, weighted scores or the JIFs annotated to publications in CVs is taking the system too far. Agrawal (2005) cites the unethical practice in which editors encourage (or expect) authors to cite articles in their journal, thus inflating their JIFs. There have also been critics of the way in which the rather opaque ranking is done by the company that does the annual journal rankings (Bohannon, 2016). Although it would seem obvious that publishing in high-impact journals is something that scientists might aspire to do, there are downsides to this practice if done too often. Some labs have even eschewed totally the notion of publishing in journals such as *Science* or *Nature* – see, for example, the Rosenbaum (2008) article titled "High-profile journals not worth the trouble." While there will always be a certain appeal when one succeeds in publishing in these journals, with acceptance rates below 10 percent one must weigh this risk and effort versus the reward.

Open Access Publishing

> Scientific publishing is a rip-off. We fund the research—it should be free.
>
> George Monbiot (2018)

Just over 40 years ago, when I started publishing scientific articles, we would mail a packet to the journal editor that included a cover letter and several paper copies of the manuscript for review. Unless summarily rejected (not sent out for review), we then waited months for a packet containing the editor's decision and feedback from the reviewers. (These packets were always opened with hopeful trepidation.) Assuming a positive response, revisions were typically made and ultimately the final manuscript would go into production, including typesetting, and the laborious task of scrutinizing the galley proofs to make sure that no typos or other errors had crept in during this process. Upon publication 6–18 months later, the hard copy issue of the journal would be mailed to libraries and individual subscribers. Needless to say, the mechanics of peer review and scientific publication has been transformed by e-communications and the internet. In addition, over the past decade, nothing has contributed more to the transformation of peer-reviewed publications than open access (OA) publishing.

In the OA model, all publications are available to anyone able to access the journal's website. The article can simply be downloaded, typically as a "PDF." Likewise, the figures can oftentimes be downloaded, sometimes even in presentation format as a PowerPoint (e.g., from *PLoS ONE*). This is a fundamentally different model from previous practice in which access to publications was available only to individual and institutional subscribers. Others could access the article online, but it was behind a "paywall" that includes a considerable one-time charge. Even if not an individual subscriber, many of us can work around these charges if our institutional libraries have e-subscriptions. But the public at large and those lacking a connection to an academic library are left out. Open access revolutionizes access. Studies have also shown that publishing in OA journals increases individual downloads and also increases citations (MacCallum & Parthasarathy, 2006), thus boosting the author's citation rate (see discussion of this topic later in the chapter). With regard to Broader Impacts, other benefits likely include greater dissemination outside of academe, such as access by K–12 teachers for instructional content.

Someone has to pay for the costs of publication. Open access shifts the burden from the journal subscribers to the author (Alizon, 2018; Van Norden, 2013). The cost to publish in an OA journal typically ranges from a low end of about $1,500 (e.g., *PLoS ONE*), to a mid-range of about $3,000 (e.g., some Elsevier journals), to $5,000 or more at the upper end (e.g., *Nature Communication* – see selected comparisons at PeerJ, 2017; also Kowaltowski & Oliveira, 2019). Some professional societies that publish peer-reviewed journals give their members discounts. In an innovative model, the recent upstart *PeerJ* offers several member categories, including the option to pay once and publish in the journal for life (PeerJ, 2017). For researchers with grant funds, paying the bill for OA may be acceptable. However, many scientists, including graduate students and researchers at smaller universities and colleges, may not have access to funds for OA publication. This model therefore favors the large, well-funded research laboratories and, in so doing, sets up a dichotomy between the haves and the have-nots. This is thus similar to the earlier model in which authors had to pay page charges, but these were oftentimes waived for those lacking funds. Such is not the case with most OA journals, although some provisions are made, such as for authors from developing countries, often on a case-by-case basis.

Established journals based on the traditional subscription model have had to adapt to catch up. Some of these journals have migrated to a hybrid model in which some articles are published in the conventional way behind the paywall, and others appear as OA in the same issue. When the journals appear online, articles can either be accessed for free (OA) or downloaded for a charge. This seems to be a temporary fix as the academic publishing industry adapts. Some funding agencies and foundations, however, see the hybrid paywall and OA model as an interim solution, with full OA as the end game (Enserink, 2018; Rabesandratana, 2018). Some recent developments from funding agencies affect the push toward OA. The National Science Foundation now requires that all research and associated data be made freely available within one year after publication (Lucibella, 2015). While formal OA is an option, NSF's policy is more inclusive and allows other means of public access, such as digital repositories, in which articles and associated data are posted on publicly available websites and servers.

There have been ripple effects as well in the world of scientific publishing. In a recent challenge to the large publishing house Elsevier, 60 German universities have banded together, refusing to pay the exorbitant journal subscriptions (Kwon, 2017; Vogel, 2017). Likewise, 100 French universities are not renewing their contract with almost 1,200 Springer journals, another expensive publisher (AAAS, 2018a). In the United States, as of March 2019, the University of California did not renew its annual contract with Elsevier, which amounts to more than $10 million for 1,500 journal subscriptions (Fox & Brainard, 2009; McKenzie, 2018). In addition to the overall subscription costs of these journals, the issue revolves around the reluctance for these large scientific publishers to move to a full OA model (Hiltzik, 2018).

It is clear that the academic community seems to be working toward a model of all funded research being made publicly available – that is, not hidden behind a paywall imposed by the publishers. The next decade should continue to see academic publishing models evolve so that research becomes more accessible within the scientific community and to the broader public. The days in which we receive our individual paper issues of journals and order "reprints" that we mail to colleagues are gone and will never return.

Preprint Services

For more than a quarter-century, physicists have been sharing early versions of their manuscripts prior to peer review. This "ecosystem" is exemplified by arXiv, started by Los Alamos National Laboratory in 1991 (Berg, 2017b). Many other STEM disciplines, such as biology, have been slow to make preprints readily accessible prior to official publication. This practice is changing, however, with the advent of services such as bioRxiv, a free biology preprint server (Kaiser, 2017a; 2017b). In an era of open communication and publication, and in response to the glacial pace of production typical in scientific publishing, preprints have value in presenting results to one's peers early. Preprints can thus promote dialog and also enhance access for colleagues and journalists, among others. Some journals, such as *PeerJ*, have developed a pipeline in which manuscripts sent to their preprint server then can go directly to the formal peer-review process. Preprints are not without challenges. In the world of systematic biology, in which new taxa are named, it remains to be seen how preprints will affect formal publication dates and priority of nomenclature. Nevertheless, just as OA is changing the landscape of scientific publishing and promoting access, so too are preprint services changing peer-review workflow and access.

Quantifying Research Outputs

The past few decades have seen increased emphasis on citation metrics of published research articles thanks to the web and automated data-mining algorithms. Two metrics are discussed here, the overall number of citations and the h-index.

The two prominent citation services are the Web of Science and Google Scholar, each with their own indexing mechanisms (De Winter et al., 2014). I typically use Google Scholar, so this method will be analyzed here. Google Scholar sums the total number of times an article is cited; this is mostly done by author, but more recently departments and programs also have these data compiled to show aggregated, unit-wide citations. Full articles in peer-reviewed journals are typically the focus, although at times abstracts and notes creep into the total citation count. Citation rankings are grouped within authors' self-selected nouns or terms, for example, "paleontology" or "evolutionary biology." For paleontology, it is unfortunate that rankings are provided separately for "paleontology" and the British English equivalent "palaeontology," when in fact these should be the same comparison group. Nevertheless, the system seems to work in most cases, and Google Scholar is considered by most a generally reliable means of providing citation metrics.

The other metric in common use is the h-index, originally proposed by Hirsch (2005). The h-index is described as the number of times a paper has been cited h times. As an example, an h-index of 35 means that 35 of an author's papers have been cited at least 35 times. The h-index was proposed to hedge the effect of an author whose overall number of citations is dominated by one, or a few, highly cited paper(s). As quoted in Ball (2005: 900), "Hirsch suggests that after 20 years in research, an h of 20 is a sign of success, and one of 40 indicates 'outstanding scientists likely to be found only at the major research laboratories'." As noted previously for the JIF, by virtue of their size or reach, different disciplines can produce disparate h values. As also with the overall number of citations, this system likely works best for comparisons and rankings within disciplines, but even as Hirsch (2005) noted, the h-index is not a fair comparison between disparate STEM fields (Technion, 2017). Other indices have been proposed, like the i-10 index devised by Google Scholar (Cornell University Library, 2018), which represents the number of papers published that have been cited at least 10 times. Nevertheless, the number of citations and Hirsch's h-index seem to be the ones most frequently used.

Like it or not, over the past several years citation numbers and the h-index have crept into tenure and promotion criteria within applicants' packets. In some fields, several hundred citations and an h-index greater than 10 might justify promotion and tenure for a junior faculty member; a total number of citations in the thousands and an h of 20–40 might be expected for promotion to a senior professor rank. One would hope that these quantitative metrics are just part of the overall assessment during tenure or promotion deliberations, and that other criteria, including grant productivity, teaching evaluations, and service, also factor into the ultimate decision.

New Ways of Tracking Research Impact

In addition to promoting yourself to your peers and science in general, anyone interested in Broader Impacts and societal benefit must also understand how to reach the general public. As has been discussed elsewhere in this book, a broader reach is not just a good idea from a philosophical point of view, but is also expected by funders like NSF. Traditionally there have been many ways in which we can promote what we do to the public; many of these are also described in more detail in the following chapters. Press releases, exhibits, and public talks have been traditional mainstays, whereas over the past few decades the development of websites, podcasts, blogs, and other social media have been a good practice for communication in the digital age. In no segment of society has promoting oneself exploded more rapidly than in the realm of social media, which provides great opportunities in the twenty-first century.

Altmetric Scores

While at an editors' meeting in January 2017 at Cambridge University Press in New York, we reviewed the performance of two journals, *Journal of Paleontology* and *Paleobiology*. In previous years we had discussed metrics such as the JIF, number of subscribers, and article backlogs (time to publication), but at this meeting I was struck by the emerging emphasis placed on the Altmetric score of individual articles published in these two journals. I was hesitant to embrace yet another fad that would pass, but it became clear that the Altmetric score was being taken seriously (Bornmann, 2014). These scores are a way of understanding how a scientific article is being disseminated and promoted to broad audiences through diverse media. Each mention by one of these media sources is given a point score, which varies from 8 for a news article to 0.25 for a Facebook post or YouTube (Inset 5.1; Altmetric, 2017). At 1.0, Twitter is given a higher score than the other social media sources. There are many other details and nuances to be learned about Altmetrics. For example, the citation system can be buffered from self-promotion – that is, multiple posts about one's own article. Similar to numbers of citations, it is difficult to compare Altmetric scores from different disciplines. For example, the highest Altmetric score recorded in 2016 was 8,063 for an article written

Altmetric score weights	
News	8
Blogs	5
Twitter	1
Facebook	0.25
Sina Weibo	1
Wikipedia	3
Policy documents	3
Q&A	0.25
F-1000/Publions/Pubyear	1
YouTube	0.25
Reddit/Pinterest	0.25
Linkedin	0.5
Open Syllabus	1
Google+	1

Inset 5.1

Obama's Altmetric

In 2016, then US President Barack Obama's article on the Affordable Care Act generated the top score of 8,063.

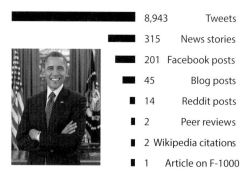

8,943	Tweets
315	News stories
201	Facebook posts
45	Blog posts
14	Reddit posts
2	Peer reviews
2	Wikipedia citations
1	Article on F-1000

Fig. 5.1 Infographic of Altmetric score for an article on US health care published by President Barack Obama (2016), modified from Altmetric (2017).

by Barack Obama titled "United States health care reform: Progress to date and next steps" (Obama, 2016; Fig. 5.1). It is of dubious significance to try to compare this to the upper end of Altmetric scores of ~ 100 for articles published in niche journals such as *Paleobiology* (Cambridge Core, 2018).

For scientists wanting to impact society, the Altmetric score should be taken seriously. On the departmental or institutional level, even more than before, it now makes sense to integrate activities that relate to Altmetric scores. Thus, press releases might be coordinated with social media, for example. It is too early to tell whether Altmetric scores will have an impact on professional scholarship, such as through a positive feedback loop to boost citation numbers. Until that time, ivory tower academics may disparage Altmetrics, but when places like large academic publishing houses (e.g., Cambridge University Press) take this metric seriously, it is time for all of us to take notice. Unlike the number of citations and the h-index, Altmetric scores have not crept into academic tenure and promotion decisions. Given the traditional emphasis on factors other than peer-reviewed research publications, this metric is likely to be of limited use in these circles. An exception, perhaps, will be in forward-thinking institutions that highly value public outreach and societal benefit as criteria for academic advancement.

Twitter, Mary Anning, and K-Index

Our seeming obsession with scholarly metrics may have gone too far. Apparently tongue-in-cheek, genomicist Hall (2014) proposed the Kardashian or K-index after the Hollywood celebrity Kim Kardashian, who is popular on Twitter. It is fascinating that in this article Hall uses the early nineteenth-century fossil collector Mary Anning as someone who contributed to natural history, but was an unsung heroine during her time. Although this makes little sense because Twitter was almost two centuries away, the author's point seems to be that had there been social media at that time, then Mary Anning would potentially have had more of an audience to promote her discoveries. This is definitely a whimsical point of view.

The Kardashian index, or K-index, is a comparison of the number of Twitter followers and scientists' citations. Hall's proposal led to a flurry of comments and criticism. Nevertheless, *Science* deemed this worthy of a note titled "The top 50 science stars of Twitter" (You, 2014). The astrophysicist Neil deGrasse Tyson has 2.4 million followers on Twitter and 151 citations, yielding a K-index of 11,129 (You, 2014). Hall (in You, 2014) mused that "Scientists with a high score on the index ... should 'get off Twitter' and write more papers." The *Science* article also notes that highly cited scientists and communication by Twitter are not necessarily incompatible.

It seems impossible to balance all of the time demands to do both science and social media, but the K-index demonstrates that it can be done if the scientist considerers public communication to be a priority. It is clear that among the various kinds of social media available today, Twitter has among the broadest reach to both professional audiences in science and the general public. Twitter also appeals to a younger demographic; 37 percent of Twitter users are between 18 and 29 years old.

Box 5.1 A Novice's Tips on Press Releases

In June 2016 we sponsored a professional development program in which K–12 teachers helped us dig at a local fossil locality in Florida. Two teachers found bones and a tusk of a six million-year-old (late Miocene) elephant-like "gomphothere" from this fossil site. Both their discovery and enthusiasm were infectious and served as a model for public participation during the dig.

Fig. 1 Teachers from Santa Cruz, California, Erin Peterson Lindberg (right) and Rebecca Mussetter (left), uncovering six million-year-old late Miocene fossils from Florida (FLMNH, Jeff Gage photo).

After the dig, I contacted Erin's and Becca's superintendent, extolling the value of their participation. I also suggested that a press release be issued about them to the local media. Kris (the superintendent) said "That's a great idea, send us one!" I was a little taken aback because I had never written a press release, and therefore did not know how to start. I proceeded by searching the internet for "how to write a press release" and a template that I could use. I filled in the blanks and sent it off to the superintendent. Soon thereafter, a reporter from the local TV station (KSBW) in California interviewed me for part of a feature they were developing for the evening news about Erin, Becca, and their fossil gomphothere discovery. This activity was featured on the evening news, likely reaching a large public audience of TV viewers in the central California region.

While many of us at universities have access to PIOs (public information officers), and we are expected to coordinate press releases with them, not everybody has these in-house resources, and our PIOs are oftentimes overextended. In the latter case, knowing how to write a press release is a good part of one's outreach repertoire, although coordination with the professional communications officers is always a good idea.

It also has a significant international reach, with almost 80 percent of users from outside the United States (Aslam, 2017). I have had a Twitter account for several years, but only sporadically contribute to it; my metrics are very low. Given the reach, particularly as a scientific means of communication, it might be time to use Twitter in a more deliberate and sustained way. Even so, I have neither calculated my K-index, nor do I have plans to do so in the foreseeable future.

Summary: Optimizing Impact in the Twenty-First Century

Collectively, scientists are a diverse group of individuals with different personalities, strengths and weakness, and motivations to promote their discipline, whether it be to their peers or more broadly to the public. Whereas in the past these practices were less structured, common sense tended to guide efforts to promote ourselves professionally. In the twenty-first century, however, the need to promote oneself to optimize impact has become increasingly more important as well as both more structured and quantified. Within our professions we are guided by metrics that quantify impact; others related to public outreach have been developed, such as the Altmetric score. The use of metrics will continue to grow in kinds and importance in the future. Bornmann (2014) discusses the potential relationship between Altmetrics, broader impact, and the overall benefit to society. Forward-thinking institutions may in the future use Altmetrics as part of the evaluation process of a scientist's effectiveness with outreach. As with all such metrics, these need to be taken into consideration as part of an overall set of evaluative criteria.

Most scientists within academic positions are expected to optimize their research impact in order to advance their careers and professional reputations. There is not enough encouragement in the reward system to do more than this, but many scientists understand the importance of broader outreach to the public. We now have at our disposal a bewildering array of tools to optimize impact, which through social media, for example, can be done simultaneously for both professional and public audiences. This trend, which has developed largely during the past few decades, is certain to continue. However, we do not know what communication tools are yet to be developed and will become available to us in the future. Promoting oneself takes time, planning, and effort; it doesn't just happen. It will always be a challenge to allocate enough time in our busy schedules to promote ourselves, while also sustaining our research activities and scholarly output.

6 Collaboration, Authorship, and Networks

Publish or Perish: A Tale of Two Advisors

In the 1970s, when I was a PhD student in the geology department at Columbia University, many of us aspired to get our first paper published in a peer-reviewed journal. A graduate student colleague of mine had worked for months on getting a paper ready, with little input from his major professor, who was a prominent scientist and apparently too busy to be involved in the student's research. When the student completed the first draft, he gave it to his major professor to read. There had been no discussion about authorship. When the paper was returned to the student, there were few comments or annotations, other than the surprising insertion of the advisor as the senior author of the paper. This seemed outrageous, and I remember us discussing his predicament. My friend fretted about what to do. After some uncomfortable discussion with his major professor, the student became senior author and his advisor was changed to the second author. This resolution was despite the fact that the advisor had contributed little, if anything, to the intellectual formulation of the paper, and he had done none of the writing.

At about the same time I was readying my first abstract for a national meeting. I likewise wondered what I was supposed to do about authorship. My major professor and I discussed aspects of my study and he gave me encouragement and suggestions for improvement. I went into his office and asked him what I was supposed to do about authorship? I recall that he responded by saying: "That is your work and you should be sole author!"

My friend ultimately changed major professors for a variety of reasons, including lack of accessibility to his advisor. It is also clear that the matter of the authorship of the peer-reviewed paper had a lasting, negative effect on their relationship. I published three peer-reviewed papers as a graduate student. I published my dissertation as sole author (MacFadden, 1977), and the other two papers were written with coauthors. My specific field of interest was somewhat different from that of my advisor. I did not coauthor a paper with him until a decade after my PhD. Throughout my career I have had great respect for my PhD advisor's academic integrity.

Introduction

A fundamental part of the process of research is collaboration and networking. The rationale and benefits of working with others are manifold. In an increasingly integrative pursuit, which at the same time has seen a growth of individual specialization, collaboration promotes the pooling of resources and sharing of diverse expertise. Many breakthroughs can come from these integrative or multidisciplinary collaborations. From a human perspective, the process of research has a social component in which scientists collaborate and make meaningful friendships, but if not careful these can also result in unpleasant consequences. Nevertheless, during my more than 40 years of research, I have largely benefitted from rewarding collaborations as well as forming lasting friendships with colleagues. Because the stakes are high in terms of professional recognition and advancement, from the beginning one should enter into new collaborations with eyes wide open and proactively set ground rules, boundaries, and expectations.

Despite public perceptions, the idea of scientists as socially awkward introverts (and older white males with beards and lab coats) toiling away in one's own individual lab on some obscure experiment is an anachronism. Collaboration is on the rise (Figs. 6.1 and 6.2). The reasons for this pattern are complex, but as science becomes increasingly specialized and also more technologically enabled, a major factor is the sharing of expertise to conduct significant inter- or multidisciplinary research geared toward breakthroughs that might not otherwise be possible as an individual.

Scientific Authorship

Publishing one's research results in peer-reviewed journals is the gold standard of scientific output and dissemination. Productivity in this arena impacts how our

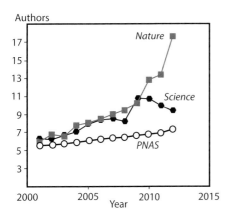

Fig. 6.1 Increase in the mean number of authors per paper in three prominent scientific journals during the twenty-first century (Persson & Glänzel, 2013).

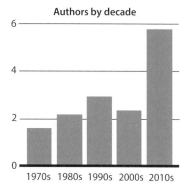

Fig. 6.2 Mean number of authors on papers that I have published, by decade. The increase from 2010 onwards is likely a result of the large multi-institutional and collaborative Panama PIRE Project . (Panama PIRE, 2017)

careers progress and ultimately how we are judged, not just by our colleagues, but in other more practical matters such as advancement and funding. It is therefore not surprising that with increased collaboration over the years (Fig. 6.1), the matter of authorship has come into focus and is fraught with pitfalls and uncertainties. Berg (2018a: 961) states that "the contributions of different authors to a given paper have remained relatively opaque."

Authorship protocols and best practices are part of professional ethics, but these are rarely taught in formal settings. Nevertheless, over the past decades the National Science Foundation (NSF) has required responsible conduct and ethics training, for example, in graduate research training grants (NSF, 2010a).

According to Strange (2008), the anecdote presented at the beginning of this chapter is an example of coercion authorship, one in which a person in power (major professor) exerts undue influence over a subordinate (graduate student). The extent to which this kind of authorship occurs within science is not well known. It is likely a practice that has been tacitly accepted in some circles, but certainly not openly celebrated. Although Strange's article is intended primarily for a biomedical audience, it is generally applicable to most aspects of STEM research. This exemplary article is assigned reading for several of our graduate seminar courses. There are several forms of unethical authorship described in Box 6.1. Another that deserves mention here is "honorary," or "honorific" authorship, in which a colleague has been added to the list of authors but does not meet generally accepted criteria for authorship. Leaders in many academic circles have called for the end to honorific authorship (e.g., Alberts, 2010). Likewise, the National Academy of Sciences (2009) published a comprehensive report, *On Being a Scientist*, that discussed these matters and ethical recommendations that promote responsible authorship practices.

With regard to authorship, one test is whether the person has made a "significant contribution" to the research and has also been involved in the process of writing the manuscript. One gray area within the test of significant contribution is when a productive scholar supports a lab and its researchers, but is otherwise detached from a specific research project. The National Academy of Sciences is clear about this when they state:

Just providing the laboratory space for a project or furnishing a sample used in the research is not sufficient to be included as an author.

(NAS, 2009: 37).

Box 6.1 Authorship Abuse and Collaboration Tips

Kinds of Authorship Abuse

"Coercion authorship
Use of intimidation tactics to gain authorship … .

Honorary, guest, or gift authorship
Authorship awarded out of respect or friendship, in an attempt to curry favor and/or to give a paper a greater sense of legitimacy.

Mutual support authorship
Agreement by two or more investigators to place their names on each other's papers to give the appearance of higher productivity.

Duplication authorship
Publication of the same work in multiple journals.

Ghost authorship
Papers written by individuals who are not included as authors or acknowledged.

Denial of authorship
Publication of work carried out by others without providing them credit for their work with authorship or formal acknowledgment … ."

(Strange, 2008: C568, table 1 [largely verbatim, abridged])

Ten Simple Rules ("Tips") for a Successful Research Collaboration
1. Do not be lured into just any collaboration
2. Decide at the beginning who will work on what tasks
3. Stick to your tasks
4. Be open and honest
5. Feel respect, get respect
6. Communicate, communicate, communicate
7. Protect yourself from a collaboration that turns sour
8. Always acknowledge and cite your collaborators
9. Seek advice from experienced scientists
10. If your collaboration satisfies you, keep it going

(Vicens & Bourne, 2007)

I typically do not agree to be an author unless I have written portions of the manuscript. I also offer this opportunity to others when assembling coauthors on an article to be published. In addition, senior authorship, which we will discuss below, typically has the responsibility of doing most of the work, not just in the generation of research data, but also the preparation of the manuscript and leading it through the review process. From a mentoring point of view, if I invest significant time advising a student and share my intellectual capital, I expect to be an author, including writing section(s) of the manuscript, or at least heavy edits of multiple drafts, if necessary. I do, however, encourage PhD students to reserve the core chapter of their dissertation for sole authorship. In so doing, there is no question about the attribution of credit for the publication, which in many ways can define their intellectual creativity and contribution to scientific research early during their professional careers.

In addition to whether someone should be an author, the sequence of co-authorship is also of importance. A bibliography (publications list) in a scientist's curriculum vitae (CV) that mostly includes junior authorship has potential implications, not just for indications of research leadership (or lack thereof), but also additional scrutiny during promotion and other forms of advancement. The different forms of authorship include sole, senior, "junior," corresponding, and last. Other than sole, the relative contributions of each author to the published article are not universally agreed upon and therefore subject not just to abuse, but also to different interpretations. As mentioned above, senior authorship should simply mean that person did most of the work. Authors other than senior authors are sometimes called junior authors, which implies a relative contribution less than the senior author. In articles with more than two authors, the sequence after the senior author may require interpretation. One way in which this sequence can be determined is the relative amounts of contribution by the junior authors, although sometimes this is difficult to determine. In recent years, in some academic circles the last author is second in importance after the senior author. Thus, last author confers a distinction of someone in charge of a project or laboratory. This practice can be problematic because some reference lists are truncated when cited (e.g., only the first five authors listed). As such, the last author in a paper with many authors is sometimes left out, and therefore lost from being recognized. This scheme converges with concerns about honorary authorship, such as in the NAS (2009) quote above. In some circles corresponding author confers a special status (Strange, 2008), although this is not the case in most of the journals in my field. With regard to papers with two authors, sometimes a statement is made that "both authors contributed equally to this paper," although this attribution is typically lost to indexing services and others who do not read the article.

Many journals are aware of author abuse and have enacted measures to mitigate this unethical misconduct. Journals such as the PNAS (*Proceedings of the National Academy of Sciences*) require that the kind and degree of participation of each author during the research development and manuscript preparation phases are explicitly described (e.g., McNutt et al., 2018). This disclosure typically occurs during the submission process, although in some journals author contributions are listed in the published article.

For many aspiring and ambitious scholars, the number (quantity) of peer-reviewed articles is considered a metric of success. The desire for increased citations can also promote abuse through reciprocal authorship ("you put me on yours and I'll put you on mine"). While this is likely not explicit, it can be implied during research project collaborations. Decades ago, NSF expected proposers to include full CVs in their grant proposals, but the numbers game of publications (quantity) has given way over the past few decades to an abbreviated (two-page) biographical sketch. With regard to the number of publications, now only 10 are allowed – five that are closely related to the proposed research and five other publications. For the first five publications, it is best practice to select those in which the principal investigator (PI) is sole or senior author, published in the most well-respected journals, and/or are most highly cited in, for example, Google Scholar. Thus, quality is emphasized over quantity. In recent years, NSF has provided more flexibility for the second five publications, and now even allows, in addition to the gold standard of peer-reviewed publications, other products such as an exhibit, video, website or other deliverable. Related to this change, another recently created section in NSF CVs is termed "Synergistic Activities." This section also provides a place to highlight other scholarly pursuits that may not have resulted in a product, but describes an activity such as work on an exhibit committee or teacher professional development (NSF, 2016b). Thus, in addition to an emphasis on the quality of representative publications, NSF fosters and encourages Broader Impacts-like activities via the second set of five publications or other products.

Clarification of authorship and the authors' respective contributions is on the horizon. McNutt et al. (2018) describe recent attempts to standardize the recognition of research contributions as they relate to authorship, using the Contributor Roles Taxonomy (CRediT) system. Hopefully the widespread adoption of this system will mitigate what has, up to now, been an intractable challenge to the scientific integrity and transparency of authorship.

Research Collaborations

Much of the discussion presented here about coauthorship is integral to understanding the practice of professional collaboration. In recent years the emphasis on collaboration has increased, mostly as a result of the importance of sharing expertise. The National Science Foundation also understands the value of the adage that "the sum is greater than the individual parts" and increasingly has developed collaborative opportunities and, as will be discussed in the following, research networks. When I started writing NSF proposals in the mid-1970s, many of the grants available at that time were geared toward single-investigator awards. My first collaborative grant from NSF was awarded in 1978 with a colleague in my department and my first multi-institutional grant was awarded in 1980. Coauthorship on peer-reviewed publications can be viewed as a proxy for the extent of collaborations. More than half of

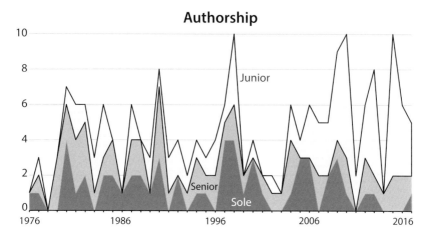

Fig. 6.3 Distribution of sole-, senior-, and junior-authored peer-reviewed publications during the author's career.

my peer-reviewed articles represent collaborations, including senior or junior authorship (Fig. 6.3).

As many others have noted, collaborations can be very rewarding, but also challenging. Vicens and Bourne (2007) provide 10 helpful tips for collaboration (Box 6.1). In all collaborations, one needs to understand the mutual benefit of the partnership from the beginning. What are you and your collaborator going to get out of this collaboration? If it is lopsided – in which the benefit does not accrue equally – then the collaboration will not likely persist without issues. This is particularly important when embarking on a new collaboration with someone that you do not know and certainly have not collaborated with in the past. Although it is awkward, candid discussions at the beginning about such matters as resources, schedule, and authorship will avert more painful communications later. Even with eyes wide open, entering into a new collaboration represents a risk – including different styles and potential outcomes.

Several years ago I received a "cold-call" email from a postdoc at another institution who had been recommended to contact me about the development of a new kind of direct dating method for fossil teeth. After numerous emails and Skype communications, we developed a research design and selected the samples for analysis. I also discussed expectations of authorship. I supplied the samples for analysis, and otherwise contributed by advising the postdoc, who was not a paleontologist. I also was prepared to write parts of the manuscript that would result from the analysis. Months passed, and we periodically kept in contact via email. When the analyses were completed, in my mind, the results were equivocal. The postdoc, on the other hand, while appreciating that the results were less than we had hoped, felt that they were nevertheless worthy of a note published in a peer-reviewed journal. I disagreed and released the postdoc from any coauthorship obligation with me. Although the collaboration did not end the way I had hoped – with a breakthrough paper in a high-profile journal – in my mind we parted on amicable terms. I also feel that if the collaboration were rekindled, perhaps resulting from new analyses, we could enter back into the partnership without prejudice.

The number of collaborations possible are finite; there are only so many that can be entered into during one's career. While at the beginning of one's career, a young researcher might enter into a more risky collaboration for a variety of reasons, more experienced senior researchers potentially have the opportunity to pick and choose their collaborations. After five decades as a scientist in academia, I have been involved in collaborations ranging from fully rewarding to intensely challenging. Being a scientist is hard enough these days, and I completely agree with Vicens and Bourne (2007) when their first rule for collaboration is "Do not be lured into just any collaboration" (Box 6.1).

Professional Networks

Many scientists realize that the beginning of the process of collaboration includes professional networking. Thus, a fundamentally important activity of professional meetings is not just the sharing of research results via talks and posters, but also the opportunity to form and sustain collaborations through informal networking and related social events. In fact, according to Aiken (2006), almost half of participants expect to develop collaborations as an outcome of professional meetings.

A large body of learning research and theory exists about the process of forming, cultivating, and sustaining learning networks, or a "community of practice" (CoP). This term, originally coined by Lave and Wenger (1991), has been widely used not just in education or science, but also in other pursuits such as economics and the business sector (Wenger et al., 2002). This concept is also relevant in the current context. In a CoP, like-minded people typically come together to develop and learn about domains of shared experience and interest. Of particular relevance here, the National Alliance for Broader Impacts (NABI; Box 6.2) is a CoP that promotes Broader Impacts among interested stakeholders.

The CoP was originally conceived prior to the advent of cyberenabled communities, but over the past few decades the mode of communication has changed dramatically. The theoretical framework of the CoP includes a defined path, process, and structure of organization. In order for the CoP to be sustained, the interest in foundational core, or content (subject) learning, must be a central focus (Wenger et al., 2002).

With regard to informal STEM learning, citizen scientists in some domains are well organized into what can be described as CoPs. For example, in the United States "star-gazers" and "bird-watchers" are organized into thematically focused groups or organizations that promote learning and sharing of knowledge related to, respectively, astronomy and ornithology. Furthermore, these domains and their participants are well organized, or "federated," again via organizations such as, respectively, the American Astronomical Society of the Pacific (2017) and the National Audubon Society (2017). These and other organizations harness the expertise

Box 6.2 NABI: National Alliance for Broader Impacts

The National Science Foundation's Broader Impacts were released in 1997 as part of the merit review criteria. Ever since that time, scientists and research administrators have attempted to find common ground among colleagues, as well as sharing mutual interests in promoting and increasing success at NSF via meaningful Broader Impacts activities. Since 2013, several initiatives have been funded via NSF to advance Broader Impacts activities via the formation of a research network, or a CoP. NABI, as this CoP is currently envisioned, is dedicated to promoting and connecting like-minded individuals interested in Broader Impacts. The goals of NABI are fulfilled by resources on their website (NABI, 2017), a listserv, an emerging social media presence, and sponsoring an annual meeting.

Fig. 1 Featured participants at the 2014 Broader Impacts Summit held in Arlington, Virginia. Left to right: Wanda Ward (NSF), Kemi Jona (Northeastern University), France Córdova (NSF), Alan Leshner (AAAS), and Susan Renoe (University of Missouri).

(Jeff Mauritzen and inphotograph.com photo)

of people with shared interests to advance domains (astronomy and ornithology, respectively, in the examples above) in the form of a CoP. An important corollary is that some of these CoPs attempt to democratize science by breaking down barriers between professionals and amateurs. In so doing, these CoPs are promoting PPSR (public participation in scientific research), which is a Broader Impacts activity of potential benefit to society.

NSF Networks: RCNs and TCNs

In 2000 NSF began funding the concept of a "research coordination network," or RCN, which typically is agency-wide, meaning that this program is offered across many of the directorates. Grants issued by RCN promote interdisciplinary collaboration and networking, particularly during early phases of project development. According to a recent solicitation (NSF, 2017i):

> The goal of the RCN program is to advance a field or create new directions in research or education by supporting groups of investigators to communicate and coordinate their research, training and educational activities across disciplinary, organizational, geographic and international boundaries.

These RCNs typically focus on broad research themes or questions within STEM, also including the development of novel technologies. Representative activities can be varied but typically funds are provided for travel and meetings and workshops in which new collaborations are formed and novel research themes are pursued. Porter et al. (2012) analyzed a sample of RCNs in the BIO (Biological Sciences) Directorate and highlights the positive benefits accrued via these collaborations, and how they can potentially affect scholarly impact. In particular, they found that peer-reviewed papers by RCN collaborations tend to be published in high-impact journals.

Since the inception of RCNs, there have been variations upon the general theme and intent of networks. For example, within the ADBC (Advancing Digitization of Biological Collections) program in the BIO Directorate (NSF, 2017j), the concept of the "Thematic Collections Network" (TCN) has developed over the past decade. The participants in a TCN collaborate on a common natural history collections discipline. Digitization activities are prioritized on specimens and collections that have the greatest potential to support future research breakthroughs. These TCNs are typically larger than RCNs, with budgets in the few to several million-dollar range, and typically have durations of up to 4–5 years. In contrast to the incubator concept of RCNs, TCNs are expected to form multi-institutional collaborations that share common goals, such as to digitize essentially all of the fossil insects of research quality in US collections ("Fossil Insect Collaborative"; iDigPaleo, 2017). Working within a consortium of 16 institutions in the United States, the oVert TCN has the goal of 3D scanning representative examples (Fig. 6.4) of 20,000 species, or 80 percent of the known vertebrate genera (Cross, 2017). These 3D images will be uploaded to Morphosource (2018), an existing digital repository for natural history and related kinds of specimens. Although primarily intended for research, these images are also being used for other purposes, including K–12 education and outreach.

From 2011 to 2019, 23 TCNs involving more than 400 US museums and 700 collections have been funded within this national digitization ADBC effort (iDigBio, 2018), which has an overall estimated budget of about $100 million over the decade-long duration of the program. These TCNs are coordinated by a national hub, iDigBio (located at the University of Florida). This hub provides infrastructure support, promotes research, provides workforce training, and coordinates education, diversity, and outreach activities (Page et al., 2015).

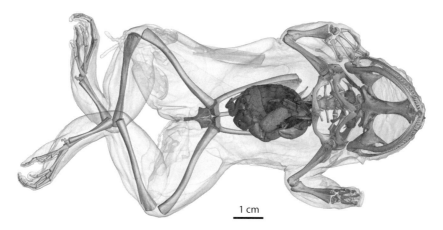

1 cm

Fig. 6.4 Virtual and computer-enhanced dissection of the Philippine flat-headed frog *Barbourula busuangensis*. This micro-CT scan (UF FLMNH Herpetology specimen 70546) was produced for the oVert (2017) TCN project. Utilizing non-invasive X-ray scanning and post-processing imaging, three-dimensional details of skeletal and soft anatomy can be shown. Also included here is this individual's last meal – the ingested crab in the middle of this image. 3D models of this specimen are also available from Morphosource (2018; Ed Stanley graphic).

Changing Landscape of Twenty-First Century Networks

If all of the ways above were not enough to keep scientists connected, the advent of social media and web-enabled connectivity are also available to develop professional contacts and networks. Of these, LinkedIn, ResearchGate, and ORCID are discussed below.

LinkedIn is a social networking platform for any professional, not just in STEM, but in any field. The purpose of LinkedIn is not so much connecting to others in your field, although that works as well, but actually anyone who desires more opportunities from making and maintaining professional connections. In a sense, LinkedIn has many of the capabilities that are also found in Facebook.

In contrast to the broad professional audience served via LinkedIn, ResearchGate is focused mostly as a social platform for academic researchers, including STEM, but essentially all other domains as well in which knowledge is produced. Kintisch (2014) likened ResearchGate to Facebook for science. ResearchGate also provides static content about one's research accomplishments and impact. For example, articles can be posted online in ResearchGate and statistics about one's research are also available on this platform. ResearchGate has developed a metric, called the RG Score, that is included in a researcher's profile and is touted as "a metric that measures scientific reputation based on how all of your research is received by your peers" (ResearchGate, 2017). It is unclear how the RG score compares to, or is better than, for example, total citations in Google Scholar or the h-index. Nevertheless, ResearchGate serves about five million academics, and is second only to Google Scholar in terms of its reach (Van Norden, 2014), although Kintisch (2014) questions whether it is "worth your time."

ORCID is a platform that provides a persistent digital identifier that links to your research and publications. A colleague recently termed ORCID a social security number for professionals that stays with them for their career. ORCID (2017) states that it is a nonprofit, global community within the global research "ecosystem."

In closing, these and other cyberenabled social media platforms are of potential value to advance professional networking and collaboration. Although I do not use many of these tools myself, I do understand their benefits. With the multifarious demands on one's time, it is indeed a challenge for scientists to stay fully connected in modern society.

Concluding Comments: Best Practices and Sustainability

Research collaborations can be both rewarding and productive, and challenging at the same time. The more complex they are, the more challenging they can be. Sometimes research collaborations can change and are no longer mutually benefi- cial. It is important to know the signs of a good collaboration (e.g., Box 6.1), while, on the other hand, to also know when it is time to move on from a collaboration that has run its course or is no longer positive and beneficial. Likewise, the gold standard of collaborations – coauthorship – is of prime importance during scientists' careers, yet procedures for entering into these collaborations are likewise potentially com- plex, particularly when many authors are involved.

Another widespread challenge of funded collaborations and networks is what happens to the project and its activities after the funding ends? We will discuss the matter of sustainability in Chapter 17. Nevertheless, of relevance here, an immense literature exists on the theory of how professional networks form, are developed, and sustained. Everett (2011) explains the difference between a hub-centered network versus a more balanced network. During the early phases of networks, information (knowledge) flows and connections are focused unidirectionally out from the hub (Fig. 6.5), but as a network matures, information (knowledge) flow is more net- worked, with multidirectional connections, including the hub and nodes. In a sustainable model, when funding is removed – for example, for the coordinating hub – then the network and collaborations continue. If the network does not evolve from the hub-centered to the more balanced model, then the network is potentially unstable, particularly after funding ends.

Thus, evidence of the distributed network, with nodes acting independently from the hub, should be carefully identified and tracked because these potentially are signs of sustainability, for example, after external project funding has ended. Collaborations and networks are vital to the advancement of STEM in the twenty- first century. As scientists develop their careers, they need to clearly understand how

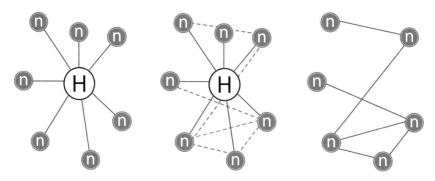

Fig. 6.5 Three phases of a developing and then sustainable network. Left: hub (H) centered network with nodes (n). Center: distributed network develops between hub and nodes and node to node (dashed line). Right: sustainable network, with the absence of the hub and one node that did not make a node–node connection (modified from Everett, 2011).

these means of collaborations and social communication can enhance their research and further contribute to their success. As mentioned above, collaborations and networks are both rewarding and challenging. Learning how to navigate these effectively is a skill that will typically result in positive outcomes during scientists' careers.

7　Strategic versus Curiosity Science

of, the & and

When I finished my PhD in 1976 I had a few copies bound and put one in the family library at my parents' home. This library consisted of just a few shelves above the TV and record player in our living room. One day my father invited me to lunch to celebrate the PhD. I was the first in my family to go to college, much less receive a PhD. At lunch, dad told me that he had looked at my dissertation. Prior to that time, I had not discussed much about the specifics of what my research was about, other than rocks and fossils. I asked him what he thought about it and he jokingly noted that "the title was interesting, and I understood three words in it – of, the, and and." He then gently remarked that he wondered what I would do in a career in which I was not a doctor or curing cancer. I should also mention that this topic never came up again and both of my parents were always immensely supportive of my career choice.

My father, a jazz musician, was a son of immigrants and grew up during the Great Depression and then during World War II. His father was a bricklayer and struggled to make ends meet. Dad was an intensely caring person who wanted the best for his sons. I am certain that in the final analysis he wanted me to be happy, but also financially successful and have a profession that was of value to society. I suspect that my story is not unique among scientists who choose to pursue esoteric topics during their careers.

Introduction

In Chapter 2 we learned about the concept of basic (or pure) research within the context of Vannevar Bush and the development of science and technology in the United States after World War II. In a broader context, the distinction is oftentimes made between basic and applied research (Inset 7.1; Prochaska, 2018). This interest in science also set the agenda for the support of research by the federal government. Of the nearly two dozen federal agencies that support either basic or applied research, only one – the National Science Foundation (NSF) – has a broad mission in all basic science and STEM fields. In contrast, the US Department of Agriculture is focused on research applied to agriculture, whereas NSF's agenda is to support basic research for its own sake. After NSF was founded in 1950, for the first few years this agency also supported medical research, but this part of its mission was transferred to NIH (the National Institutes of Health). As such, NSF no longer funds projects that are specifically focused on medical research applied to human health.

The National Science Foundation has largely preserved its focus on basic research. Nevertheless, over the years there have been several attempts to incorporate applied research within the agency – for example, the RANN (Research Applied to National Needs) program resulting from the Daddario–Kennedy amendment in 1968 (see Chapter 2). These applied science initiatives typically have been ephemeral within NSF; after its inception in 1969, the RANN program died in 1977 (Mazuzan, 1994). Thus, the core mission of basic research has been the hallmark of NSF since its beginning. As a consequence, however, the importance of societal benefit ("welfare"), and more recently Broader Impacts, has been woven into the fabric of NSF, either implicitly or explicitly.

> **Basic versus applied research**
>
> **Basic** research
> (also *fundamental, pure*, or *curosity*) is driven by a scientist's curiosity or interest in a scientific question. The main motivation is to expand knowledge, not to create something. An example is "How did the universe begin?"
>
> **Applied** (or strategic) research is designed to solve practical problems of the modern world, rather than acquire knowledge for knowledge's sake. The goal is to improve the human condition. An example is the quest to cure cancer.

Inset 7.1

In this chapter we discuss the differences between curiosity and strategic research, charismatic STEM components or disciplines, and strategic trends within NSF. A better understanding of these topics will likely help to understand not just how NSF works, but also how to optimize one's research agenda and do a better job of Broader Impacts.

Curiosity versus Strategic Research

A variant of pure versus applied science is the concept of "curiosity" versus "strategic" research. I use this as an example in my Broader Impacts graduate seminar class. Curiosity research is done because you are interested in it yourself. I might be fascinated about describing a new extinct species of fossil horse, or curious about better understanding when they lived in the past. However, this curiosity or interest is irrespective of any concern over whether this topic might be of interest to others, not just within my profession, but in a broader context within science or society at large.

On the other hand, strategic science (or STEM) is the choice of a research topic with the understanding that it fits into a broader predefined research agenda. Thus, my curiosity in an extinct species of fossil horse might also fit into a broader research agenda. In a modern context, two topics of strategic, or societal, relevance are evolution and climate change. If properly framed, fossils horses can address both of these and contribute to strategic science. Thinking more broadly, both of these topics are important for K–12 STEM education as well as public understanding of science.

In discussions with students, I hasten to add that there is nothing wrong with doing curiosity-based science, particularly if you are fortunate enough to be paid to do it and can find the funds necessary to support the research. However, if you can fit your research, or the questions that you ask, within a framework of strategic science topics, then you likely will broaden the reach of your discoveries as well as have more chances for funding opportunities.

Charismatic Research

Certain topics within science, and more broadly STEM, are simply intrinsically more interesting, or charismatic to the general public, particularly the 40 percent of our population who are curious about science and related topics (Britt, 2006). Elsewhere in this book we discuss a similar topic, that of charismatic species within the discipline of paleontology. Thus dinosaurs, giant extinct sharks (e.g., Megalodon), and human fossils all require little introduction. Other topics are charismatic as well, such as large animals and plants (the so-called "P. T. Barnum effect"), astronomical phenomena such as eclipses, and more recently drones and 3D-printing in many facets of STEM. There are, however, many more examples and it is up to the individual scientist to find charismatic topics to study if they want to drive the appeal of their research. To eschew this potential strategy is myopic. Thus, in the long run of a career, it is far easier and of potentially more societal benefit to study and promote one's research if it is fundamentally charismatic.

Strategic Trends at NSF

Core Funding Elements

Since its inception in 1950, and with the first cycle of funded proposals in 1952, NSF has supported the fundamental elements of basic research and workforce development. The original research included basic science, technology, and medical research, but the latter theme was removed from NSF's mission a few years later (also see discussion in Chapter 2). With regard to workforce development, the graduate fellowship programs started with the first funding cycle in 1952 and have been a priority ever since. The so-called GRF (Graduate Research Fellowships) is the longest-standing program in NSF and will likely continue to support future scientists from diverse fields and backgrounds. Workforce development was an essential element in *Science: The Endless Frontier* (Bush, 1945), and this was codified in the National Science Foundation Act of 1950 (Public Law 81-507, 1950). Another essential and enduring component of the NSF vision has been the importance of international collaborations, which likewise has its origins in *Science: The Endless Frontier*, the NSF Act, and strategic decisions made by NSF ever since. For example, in addition to the disciplinary research directorates (R&RAs) and education, NSF continues to support a wide range of international activities through the OISE (Office of International Science and Engineering). Our Panama PIRE Project (Panama PIRE, 2017) was funded by NSF's OISE.

Another important aspect of science is the research infrastructure that enables science and STEM in the United States. This is evident in numerous programs and initiatives that have included one-time support, for example to purchase equipment during the IGY (International Geophysical Year; see the next section) and the more recent long-standing MRI (Major Research Instrumentation; NSF 2017k) program. In the mid-1960s a short-lived program was initiated at NSF that helped with "bricks-and-mortar," the construction of buildings and facilities (our museum building [Fig. 2.4] received NSF support in 1966; MacFadden, 2017). Physical research infrastructure, such as buildings, has for the most part been within the purview of the host institution. As such, this commitment on the part of host institutions justifies the allocation of overheads as part of the indirect costs included in NSF budget requests. In addition to one-time allocations for research infrastructure such as research vessels (see discussion of ARRA in the next section), NSF also has long-term commitments for ongoing major research infrastructure, for example large telescopes and synchrotrons. It is clear that in addition to short-term investigator-focused awards and one-time infusion of funds, for example with the MRI program, NSF understands the strategic value of long-term investments in research.

Special Programs and Strategic Initiatives

In addition to the breadth of research activities supported by its myriad programs, NSF periodically promotes themed programs of limited duration. Some of these have lasting impacts, not just during a limited time period, but also resulting from the infrastructure and discoveries made thereafter. Two such programs are described here.

International Geophysical Year (IGY). In the 1950s, soon after the founding of NSF, the United States became involved in "big science" as part of a global initiative involving about 50 countries. This was the first time that such a large multinational program focused on a common theme had been undertaken for basic science. The IGY, which officially lasted from 1 July 1957 until 31 December 1958, was dedicated to advancing understanding of the Earth's oceans, atmosphere, mantle, and the cosmic influences of the sun and other extraterrestrial sources. The IGY also focused on the polar regions of the Earth; lasting effects of this initiative are seen in the Office of Polar Programs within the Directorate for Geosciences (GEO). Overall, $39 million was appropriated in 1956 and this was spent largely on individual proposal awards in 1957 and 1958 (Mazuzan, 1994; NSF Annual Reports, 1956–1958). IGY served as a successful model for "big science" as part of the overall NSF portfolio that continues to the present day. These large, interdisciplinary, and multinational initiatives are recognized to have the potential to make major breakthroughs in science (Berg, 2017a).

American Recovery and Reinvestment Act (ARRA). The ARRA was enacted in February 2009 by the 111th Congress during the Obama administration to provide stimulus for recovery from the Great Recession. The intent of the ARRA was to fund "shovel-ready" projects and get people back to work, and the funds were spread among different federal agencies to disburse. Of the ~$800 billion appropriation, $3 billion went to NSF during financial year (FY) 2009 and 2010. This additional infusion of funds represented a considerable, albeit temporary, boost to the original $6.5 billion appropriation for FY 2009 (Fig. 2.5). This was used to fund almost 10,000 projects in the research (R&RA) directorates. Befitting the mission to support people and the workforce, in education (EHR), ARRA funds went toward several programs, including Noyce Scholarships for undergraduate pre-service teacher training, Math and Science Partnerships (MSP), and the Science Masters Program (SMP). Three large, presumably close to "shovel-ready" research infrastructure projects were funded, including an Alaskan research vessel, solar telescope, and ocean observation initiatives (NSF, 2010b).

As a new program officer in July 2009 (the "summer from hell" for overworked NSF program officers; Mervis, 2009), I observed the impact of the infusion of ARRA funds into my program (Informal Science Education). A supplemental allocation as part of the ARRA funds (Fig. 2.5) to NSF came through to our program. It allowed funding of some additional highly ranked projects, and in particular those that promoted climate change education and literacy. This process aligned with the director of the NSF, Arden Bement's, strategy to help remove the backlog of highly ranked proposals; in fact, the proposal success rate – that is, those that were funded – jumped from 25 percent in 2008 (before ARRA) to 32 percent in 2009 (Mervis, 2009; NSF, 2009b). The ARRA allocation to NSF perhaps did not have the overall impact of the more deliberate strategy of the IGY. Nevertheless, it made a difference in the total number of projects funded and also fit with the overall criteria of shovel-ready projects and getting people back to work.

Recent Strategic Initiatives and Opportunities at NSF

The National Science Foundation is an organization that changes in response to input from numerous sources, including Congress, the National Science Board (NSB), and STEM community. Following these leads, NSF typically organizes internal working groups, consisting mostly of program officers, that result in the development of new proposal solicitations. These new programs oftentimes focus on strategic trends that are agency-wide. They typically have a finite lifetime; most NSF programs last about a decade unless they are very successful, or fit into the core mission of the foundation, such as the GRFP that has been part of NSF since its inception. Some examples of recent strategic investments in STEM are as follows.

"Cyber-." This prefix has been used in several contexts over the past half-century and is now engrained in many aspects of modern society (Inset 7.2; Merriam-Webster, 2017). "Cyberspace," "cyberenabled," and "cyberlearning" are examples of terms used to indicate computer-assisted activities or environments. More recently, the concept of cybersecurity has become of increasing interest (e.g., NSF, 2017l). The National Science Foundation has an entire directorate – CISE (Computer and Information Science and Engineering) – devoted to advancing research and education on computers and related technology. I have

> **"Cyber" prefix meaning**
>
> of, relating to, or involving computers or computer networks (e.g., internet)
>
> Example: *cyber*security

Inset 7.2

been involved in several "cyberprojects" funded by NSF over the past several decades. For example, the "Fossil Horses in Cyberspace" (FHC) project was started two decades ago – a time when many people accessed the internet via dial-up telephone modems. When I mention this technology to my students, they look at me like I am from another planet. I have also been advised by them and others that, in the early twenty-first century, "cyberspace" is passé.

Funded in 1996, the FHC was a supplement to an existing research grant on fossil horses. This program was led by Barbara Butler, a program officer in Informal Science Education. Her vision was to develop a grants program that bridged the research and education directorates. The FHC project was one of a few funded in the initial cohort of this program, which ultimately morphed into Communicating Research to Public Audiences. (I managed the latter program 14 years later as a program officer in Informal Science Education.)

"Fossil Horses in Cyberspace" was intended to be a cyberexhibit experiment. Given the technology of the time, and prior to the concept of an interactive Web 2.0 (O'Reilly, 2005), websites and communication flow were primarily one-way, with information flowing from the web to the user. Blogs and other social media were either in their infancy or did not exist before Web 2.0. Despite its antiquity, two decades later traffic on FHC remains heavy; this site surpassed expectations when it was created. In hindsight, it would have been informative to track and evaluate the use of FHC. Anecdotally, based on periodic email requests, FHC has been used by K–12 teachers, for science fair projects, and in books and other publications dealing with evolution. Our IT office at the Florida Museum of Natural History reports that through Google Analytics tracking, despite its static and outmoded design, FHC remains within the top 10 (of almost 20,000) pages accessed on our website, with more than 90,000 visits during the first quarter of 2018. While this is only a modest metric of traffic in contrast to some highly visited websites with educational content, integrated over 20 years FHC has had a long-standing presence and impact on the web. FHC remains one of the top returns if one searches the web for "fossil horses" (Fig. 7.1).

Hot-button topics: evolution and climate change. As discussed in Chapter 1, several topics in modern society are considered controversial or "hot-button topics" (Leshner, 2010). With regard to NSF's mission, evolution and climate change are relevant. Evolution is a theme that cuts across several directorates, including BIO, EHR, GEO, and SBE (Social, Behavioral, and Economic Sciences). Over the past several decades few, if any, initiatives have brought evolution to the forefront (such as an agency-wide

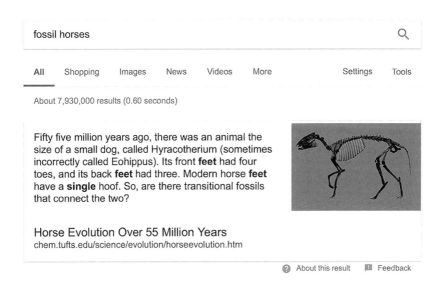

fossil horses   🔍

All   Shopping   Images   News   Videos   More   Settings   Tools

About 7,930,000 results (0.60 seconds)

Fifty five million years ago, there was an animal the size of a small dog, called Hyracotherium (sometimes incorrectly called Eohippus). Its front **feet** had four toes, and its back **feet** had three. Modern horse **feet** have a **single** hoof. So, are there transitional fossils that connect the two?

Horse Evolution Over 55 Million Years
chem.tufts.edu/science/evolution/horseevolution.htm

    ❓ About this result   🚩 Feedback

Fossil Horses in Cyberspace - Florida Museum of Natural History
https://www.floridamuseum.ufl.edu/fhc/ ▾
The Florida Museum of Natural History presents this on-line exhibit on **fossil horse** (Equidae) paleontology, evolution, phylogeny, relationships, biology, ...

Fig. 7.1 Using an internet search engine for the phrase "fossil horses" returns the *Fossil Horses in Cyberspace* website. The horse evolution page above was developed by Tufts University based on the FHC website, shown below. (The skeleton image is updated from the original.)

solicitation), but this topic is part of the underlying fabric of NSF. Many of my projects funded by NSF have evolution as the fundamental theme ingrained within them (e.g., fossil horse evolution), or they were proposed specifically to promote this topic (FHC). Indeed, any paleontologist or biologist looking for funding at NSF will likely include evolution in some aspect of potentially fundable projects.

Likewise, since the beginning of NSF, climate science or the modern buzzwords of "global change" or "climate change" have been engrained in the missions of several of the directorates. In contrast to evolution, however, which has been pervasive but typically not the focus of strategic initiatives within NSF, since 2009 (NSF, 2009c) climate change has been a priority of several initiatives, including the decade-long CCEP (Climate Change Education Partnership). It is unfortunate that the existence of global warming (a type of climate change) has become a politicized, and therefore controversial, topic within US society (Leiserowitz et al., 2017). The Pew Research Center reports that trust in climate science is low among Republicans and considerably higher among liberal Democrats (Funk & Kennedy, 2016). It therefore remains to be seen the extent to which climate change, climate science, and climate education will remain in the forefront of strategic initiatives at NSF.

"Big Data." Advances in computer architecture and infrastructure have resulted in major advances in rapid data analysis and storage. As a result of these innovations and advances, previously intractable research questions are now potentially solvable with the advent of "Big Data" (NSF, 2015; Inset 7.3), also termed data science. The production of data has always been part of the scientific process, but the generation of massive quantities has skyrocketed exponentially over the past few decades. Aligning with this trend, NSF has both understood and promoted the development of data science. For example, several solicitations in gradu-ate education (e.g., NSF, 2017m) specifi-cally focus on projects that involve big data. In fact, the field of bioinformatics has grown out of the big data revolution resulting in entirely new careers for STEM professionals entering the workforce (Levine, 2014).

> **Big Data @ NSF**
>
> NSF is a leader in supporting Big Data research efforts. These efforts are part of a larger portfolio of Data Science activities. NSF initiatives in Big Data and data science encompass research, cyberinfrastructure, education and training, and community building.

Inset 7.3

Within the field of natural history, and of relevance to biodiversity and system-atics, NSF has invested over $100 million to "digitize" museum voucher specimens in the United States through its ADBC (Advancing Digitization of Biodiversity Collections) (NSF, 2017j; Fig. 7.2). Coordinated by the iDigBio project (Page et al., 2015; iDigBio, 2018), this initiative has involved more than 700 research collections in 400 museums and has now aggregated more than 100 million specimen data records. It includes hundreds, if not thousands, of professors, researchers, museum and computer professionals, and university students in this national digitization effort (Fig. 7.2). Prior to this initiative, if a researcher wanted to obtain natural history specimen data, the curator in charge of the collection would need to be contacted for data on their desktop computer, or query individual institutional

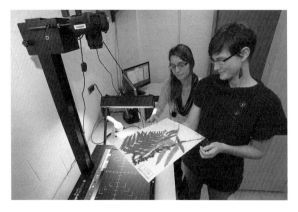

Fig. 7.2 Undergraduate students Emilie Jordao (left) and Abigail Hollingsworth (right) digitizing plant specimen (cinnamon fern) in the Central Michigan University Herbarium (Anna Monfils photo).

databases online. There was no mechanism to access all relevant data in a "one-stop shop." This has changed dramatically with the advent of big biodiversity databases such as iDigBio (2018) and the Global Biodiversity Information Facility (GBIF, 2018), for the paleontological world, the Paleobiology Database (PBDB, 2018). Thousands of peer-reviewed papers have been written using these massive databases (GBIF, 2018), and are a testimony to the fact that questions previously unknowable are now able to be answered. The applications of big data to paleontology are likewise enormous, and research enabled by these databases is increasing exponentially (e.g., MacFadden & Guralnick, 2016).

Many unresolved questions persist in science. A seemingly intractable controversy in paleontology has been the explanation for how and why 50 species of large mammals ("megafauna"), such as mammoths (*Mammuthus*), mastodons (*Mammut*), giant bison (*Bison*), and horse (*Equus*) that inhabited North America during the Pleistocene Ice Ages became extinct. Explanations for the "late Pleistocene Megafaunal Extinction" have typically focused on either climate change or human influences, the latter of which has been classically termed the "Pleistocene Overkill Hypothesis" (e.g., Martin, 1984), although direct evidence for temporal overlap and potential coexistence has been meager or equivocal (Fig. 7.3). This debate continues, with some authors advocating overlap of humans and extinct megafauna (Barnosky et al., 2004; Johnson et al., 2013), whereas others conclude that temporal overlap did not exist (Lima-Ribeiro & Felizola Diniz-Filho, 2013). It is possible that new statistical techniques (Emery-Wetherell et al., 2017), coupled with larger, temporally calibrated occurrences of these extinctions for different taxa, particularly for

Fig. 7.3 Mural depicting humans hunting megafauna in Florida during the late Pleistocene (FLMNH, Jay Matternus artist).

the abundant record in North America, will clarify the debate about the cause of these megafaunal extinctions during the late Pleistocene.

INCLUDES

Ever since Bush's (1945) *Science: The Endless Frontier*, and NSF's beginnings in 1950, science leaders have understood the importance of contributions by people of diverse backgrounds. This is exemplified by, for example, the first NSB, which included women and African-Americans (Fig. 2.2). Therefore, recognition of the contributions of diversity, not just to the scientific enterprise, but to the STEM workforce in general, has been a fundamental core value of NSF's organizational culture and expectations. Interest in diversity, inclusion, and access for all is not a new concept. It is ingrained within Broader Impacts expectations and special cross-cutting or agency-wide programs at NSF. This core value is the central focus of the HRD (Human Resource Development) Division at NSF, which is dedicated to diversity, inclusion, and equal access. Since 2014, when France Córdova (NSF, 2017n; Inset 7.4) became director of NSF, the development of the INCLUDES program has further underscored the importance of all Americans participating in the scientific enterprise in the United States (NSF, 2017n).

"Our nation's future prosperity relies on advancing the frontiers of science – and reaching our full potential requires including all Americans in that effort."
—France Córdova

Inset 7.4

Thanks to Dr. Córdova's leadership, diversity, equity, inclusion, and broadening participation have become further engrained in the core culture of NSF. In addition to supporting basic research discoveries, NSF continues its long-standing mission to develop a better-prepared workforce with people of diverse backgrounds and perspectives. We will return to this topic in Chapter 9 and also conclude the book (Chapter 19) with "Broader Impacts 3.0" – that is, the strengthening of broadening participation in the twenty-first century.

Sustainability

The concept of the Earth's finite resources and a global population that exceeds seven billion has resulted in the need for sustainable growth in the future. Thus, over the past few decades, the concept of "sustainability," or even a new field of "sustainability science," has become more of a priority for concerned members of society. The National Science Foundation takes the concern for the future of our planet seriously and has accordingly highlighted sustainability as an agency-wide initiative. The mission of the SEES (Science, Engineering, and Education for Sustainability) program is:

> To advance science, engineering, and education to inform the societal actions needed for environmental and economic sustainability and sustainable human well-being.
>
> (NSF, 2017o)

The SEES initiative thus pervades the content of programs from different directorates. Of relevance here, this theme also can be used as a type of Broader Impacts activity because of its potential benefit to society. In addition, some programs, such as the large Partnerships for International Research and Educations (PIRE) solicitations (NSF, 2011), specifically include SEES as a priority focus for that cohort of proposals.

The National Science Foundation also has embedded within part of its culture the notion that funded projects represent *investments* in research and education. The implication, therefore, is that NSF will provide resources to mobilize and support the project for a finite period of time. Thereafter, the expectation is that the research team and participating institutions will sustain the project so that it does not die after NSF funding ends. In practice, an effective sustainability plan is indeed a challenge for any project. A sense of sustainability is also now considered an important component of the review process, particularly for projects that have resulted from large NSF investments. As pressure increases for funding and project development, the importance of sustainability will only grow in the future (also see Chapter 17). This concept has become part of NSF's culture and it therefore is important for proposers to take this aspect of their project seriously.

Concluding Thoughts and the Future

With community input, NSF has developed a series of "Grand Challenges" or "Big Ideas" for the beginning of the twenty-first century (Mervis, 2016; NSF, 2017p; Inset 7.5). These are proposed in order to consolidate and prioritize strategic research agendas and themes for the next few decades. For any scientist trying to align their individual research with strategic objectives, following these 10 priorities makes sense. Even if one's particular research does not fit directly into the themes that pertain to STEM research, there are still opportunities for alignment with the infrastructure and diversity (INCLUDES) components of NSF's 10 Big Ideas.

A common strategy of NSF is a balance of: (1) long-standing programs such as GRF that are engrained within the fabric of the organization since the first grants were awarded in 1952; (2) strategic core initiatives, typically with durations of about a decade; and (3) new and flexible initiatives that respond to congressional and community input, as well as align with research needs of societal impact, for example climate change. Likewise, although STEM was not in the dictionary in the

---

10 Big Ideas for future NSF investments

1. INCLUDES: Enhancing science and engineering through diversity
2. NSF 2026
3. Understanding the rules of life: Predicting phenotypes
4. Future of work at the human-technology frontier
5. Mid-scale research infrastructure
6. Windows on the universe: The era of multi-messenger astrophysics
7. Navigating the new Arctic
8. Harnessing data for twenty-first-century century science and engineering
9. The quantum leap: Leading the next quantum revolution
10. Growing convergent research at NSF

Inset 7.5

early 1950s, STEM workforce development is fundamental to the fabric of NSF. Within this context, diversity, inclusion, and access for all continues to be a core value of NSF as it strives to lead basic science in the United States. Many of the programs described above (e.g., INCLUDES) are self-contained strategic initiatives that will likely have a finite duration. However, as these fit into the long-term organizational goals of NSF, they will likely morph into, or be followed by, equally innovative and significant programs.

8 Know Your Audience

First Things First

I started as a program officer at NSF in July 2009. Despite the fact that it was a particularly busy time for the other program officers with the influx of ARRA funds (the "summer from hell;" Mervis, 2016), I was fortunate to be mentored by an experienced colleague, Al DeSena. As a paleontologist, I came into the Education Directorate (EHR) from a science content background. In retrospect, I was clueless about how proposals in the Lifelong Learning Cluster (Informal Science Education) were supposed to be evaluated. One of the first things Al taught me was to "know your audience." Without this knowledge, it is difficult or impossible to focus and optimize education and outreach projects. This guidance served me well over the next year at NSF.

Since my time at NSF, I have learned that knowing your audience is not just something useful for evaluating proposals. In fact, this advice extends to many other aspects of what we do as scientists, including Broader Impacts activities. It is important to carefully gauge one's audience, whether it is for an academic seminar, teaching excited third-graders about fossils, determining how to write a peer-reviewed paper, or for a talk at a science café.

Introduction

Communication and learning professionals understand the critical importance of knowing your audience. Professional societies such as the Visitor Studies Association (VSA) have been created to specifically promote audience research within museums and other informal education settings. They also publish the peer-reviewed journal *Visitor Studies, Theory, Research, and Practice* that contributes knowledge to the field (VSA, 2017). This is just one example of how the professional community promotes knowing your audience; there are many more.

Scientists typically communicate with a narrow audience within the ivory tower. As such, they can use a consistent, defined vocabulary and jargon that is within their own framework of thinking and specific content domain. So too is the case when we teach university students; they are expected to learn terms and concepts that we present to them. In other settings, we oftentimes do not fully understand the intended audiences when we present outside our comfort zone, such as public talks and lectures. In fairness to the profession, however, the landscape is changing and many scientists understand the value and importance of optimizing communication once they understand their intended audience. Initiatives such as TED Talks (2017) are an example of how communication styles are changing in the twenty-first century.

The composition of an intended audience can be relatively controlled and monolithic, similar to an academic seminar at a university. In contrast, audiences of an outreach activity may be less homogeneous and therefore have a more complex demographic. From the point of view of social responsibility, but also the pragmatic aspects of NSF expectations within a Broader Impacts plan, any successful outreach activity will benefit from knowing your audience. Thus, in addition to knowing how and where you will be communicating, another fundamental aspect is to whom you will be communicating, and what knowledge and cultural backgrounds they bring with them. It should also be noted that these are not mutually exclusive. For example, a formal talk presented to a real-time audience in an auditorium might also engage learners afterwards online if the talk is digitally archived for viewing.

I was once invited to present a talk about fossil horses during a weekend fossil festival. Conscious about knowing my audience beforehand, I spoke with my host about who would be at the talk. The audience was described as: (1) the general public who were interested in fossils (and had chosen to come to the festival); and (2) local fossil club members who had considerable prior interest and knowledge about paleontology. When the talk began, from a scan of the audience I realized that some of the crowd consisted of multigenerational families, several including small kids and toddlers. As I started talking about my research on the geochemistry of horse teeth, which had previously worked well at fossil club talks, it became clear that I had missed my mark – some of the parents and small children started to leave. This kind of disruption never looks good; it was a clear indication that I had misjudged some of my audience. Nevertheless, I found some solace in the fact that others in the audience, including the members of the local fossil club, were perfectly happy to stay for my talk and listen to what I had to say. Sometimes, despite best intentions, one can miss the mark in predicting audiences.

Formal Audiences

Formal audiences are reached for a specified period of time in a structured setting, such as a classroom. This could include, for example, an academic seminar, or

Box 8.1 Donation to Children's Museum: What Would You Do?

You are the director of a small, successful children's museum in a beautiful new building with room to grow. Your primary mission is to serve young (pre-K to second-grade) learners and their families so that they better understand STEM. Your museum is located in a state with a relatively high proportion of fundamentalist Christians and creationists. One of your board members wants to fund an exhibit hall on "natural selection and evolution." Her family will pay for the entire exhibition, including the renovation of the space, design, and fabrication of the themed content, as well as funding the staff needed to develop related educational programming. This represents a substantial potential gift.

With this background, as the director of this museum, should you accept the donation to build this exhibit on natural selection and evolution? The answer to this question is "probably not." The reason for this is that evidence from learning research has shown that young children, who are the primary target audience for this museum, are not ready in their cognitive development to understand the concept of natural selection and the origin of new species through evolution. Evans (2006; 2013) has shown that young learners view the world as a series of distinct kinds (species) that do not evolve into other kinds. Thus, a hall of natural selection and evolution would not be a good fit for this museum, despite the generosity of the prospective donor. In fairness to the donor, and in the best interest of your audience, you should gracefully decline this gift because it does not fit with your institution's mission.

a presentation at a local school. It also could be either face to face or delivered online (e.g., via a webinar). Formal audiences have certain expectations depending upon their composition. Perhaps the most predictable audience is the one closer to an academic's comfort zone – the seminar or college classroom. In these settings, the style of communication is typically more presenter-focused, and for academic seminars oftentimes follows the scientific method (Fig. 4.1).

The other vast formal audience includes teachers and students in K–12 classroom settings. With more than three million teachers and nearly 51 million students in K–12 in the United States today (NCES, 2017), the potential impact on this audience is enormous and has broad benefit to society. As we will discuss in Chapter 11, there are many things to consider if one wishes to succeed in reaching K–12 audiences. To enter a K–12 classroom with little preparation or knowledge of the level of students' learning is flirting with disaster. Even within K–12 formal education, how one presents a talk to kindergarteners versus students in a high-school advanced placement (AP) biology class is an art that needs to be both understood and developed.

Fig. 8.1 University of Florida graduate student Michelle Barboza presenting a talk at the Women in Paleontology program at the Orlando Science Center in May 2017 (Rachel Narducci photo).

Informal Audiences and Lifelong Learners

In an article titled "The 95 percent solution," Falk and Dierking (2010) show that most of a person's opportunity to learn occurs outside of the classroom and throughout their lifetime. The ways in which STEM can be learned include a host of venues, including built environments (such as museums), community science initiatives, hobbyist clubs (Fig. 8.1), festivals and fairs, public spaces (e.g., airports), media such as TV and radio, and via the internet. Learning researchers have referred to informal STEM education as 'free-choice learning' (Falk & Dierking, 2002) that can occur at any time and in any place. Fifty million people visited natural history museums in 2005 (MacFadden et al., 2007); this number approximates the number of students in US schools. Natural history museums, however, are only one potential venue for informal learning; if the other ways (listed above) in which informal STEM can be communicated are aggregated, then this type of learning is at least an order of magnitude greater than formal K–12 learners in the United States today. Although more will be discussed about informal learning in Chapter 12, we need to understand that informal audiences and learning opportunities are pervasive in modern society.

Non-Formal Audiences

Typically, the dichotomy in STEM education has been between formal and informal learning. If formal education occurs in the classroom, and informal learning is free-choice, then some kinds of audiences do not fall into either of these two neat categories. Is a public lecture to a fossil club (Fig. 8.1), or a training session for museum docents, informal or formal education? The distinction between these two definitions blurs in some instances. The concept of non-formal education, which can be defined as any organized learning activity that takes place outside the formal educational system, can accommodate most exceptions to the dichotomous formal versus informal audiences. For a public audience, the speaker is well-served to flip the communication pyramid (Fig. 4.1) and discuss the overall significance or "who cares" near the beginning of the presentation in order to capture the audience's interest.

Non-formal learning typically does not include a formal curriculum or sylla-bus as is standard in formal classroom settings. How about an optional "extra credit" K–12 class field trip to a local nature center? It satisfies the informal "free-choice" criterion, but if it relates to a specific classroom lesson or curri-culum, then it also has a formal component as well. Eshach (2006) also makes the argument that non-formal education bridges the perceived gap between formal and informal environments. Likewise, the recently established journal *Connected Science Learning* (2018) seeks to integrate informal STEM and formal K–12 learning activities. Regardless of some of the overlap and apparent ambi-guity with the terms formal, informal, and non-formal audiences, it is beneficial to understand these within the context of how scientists can reach out for societal benefit.

Multicultural and International Audiences

Almost every audience will have some component, or demographic, represent-ing cultural diversity. The extent to which this exists within an audience can vary – for example, giving a talk to a Christian men's group would almost certainly be less diverse than a general public lecture at an urban science museum. The presumption so far about audiences has been that the commu-nication would be in a single language (e.g., English). However, many audi-ences are becoming increasingly diverse (Fig. 8.2). In formal K–12 education this is a persistent challenge for students who are ELL (English language learners), and also for their teachers. During my sabbatical (2015–2016) in

Museum visitors

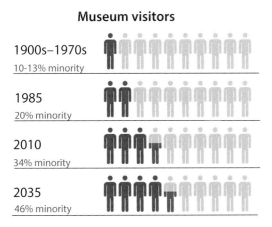

1900s–1970s
10-13% minority

1985
20% minority

2010
34% minority

2035
46% minority

Fig. 8.2 The increase in minority visitors to museums in the United States, both actual and estimated into the future (AAM, 2010).

Fig. 8.3 The *Canoes* exhibit at the Florida Museum of Natural History showing an entry graphic with bilingual content with the title Dugont Canoes (English) and Canoes Indígenas (Spanish) (FLMNH, Jeff Gage photo).

Santa Cruz County, California, I saw this first-hand with schools (e.g., DeLaveaga Elementary School) that used "two-way immersion," meaning that instruction was delivered for part of the day in English and the other part in Spanish (SCCS, 2018). I also saw, within the same county, other schools that had recent influxes of native Spanish speakers but were placed in English-only classes. This was obviously a struggle for these children. In one case, an immensely caring middle-school teacher would use her lunch period to talk in Spanish with these students about science.

With regard to informal, museum environments, another consideration is that the cultural backgrounds of audiences can change as part of evolving visitor demographics. At the Florida Museum of Natural History, two decades ago the vast majority ($\sim$99 percent) of our visitors had English as their native language. Since then, the demographics have shifted and our visitor audience now consists of about 10 percent native Spanish speakers. This trend is consistent with what is seen nationally with increasingly diverse audiences (Fig. 8.2). We therefore need to understand the best way to reach these visitors so that exhibits and public programs are accessible for them. An obvious suggestion is to have bilingual exhibits that provide content, such as text and graphic panels, in both English and Spanish (Fig. 8.3), although other strategies such as bilingual docents can also be included in best practices.

In 2013, I was invited to give a formal public lecture at the Explora science center in Medellin, Colombia, as part of the kick-off activities for a Darwin exhibit (AMNH, 2017) celebration. I predicted that the audience for this evening talk would be: (1) generally interested in science; (2) primarily free-choice, lifelong learners; and (3) largely native Spanish speakers. Given the last, I committed to

Hace cuatro millones de años se formó Panamá como resultado de la actividad geológica, creando un puente de tierra que conectó América del Norte con la del Sur. Muchos animales, tigres diente de sable, armadillos, perezosos gigantes, mastodontes y caballos... migraron entre las dos Américas. Esto provocó una mezcla entre animales nativos e inmigrantes, llamada por los científicos el Gran Intercambio Biótico Americano (GIBA). Muchos de estos mamíferos se extinguieron al final de la era del hielo, hace unos 10.000 años.

Perezoso gigante

Intercambio biótico de las Américas

NATIVOS E INMIGRANTES: LA PODEROSA MEZCLA

Invitado:
BRUCE MACFADDEN
Paleontólogo PhD, curador y profesor de paleontología de vertebrados en el Museo de Historia Natural de la U. de Florida. Estudia mamíferos extintos y ha realizado investigaciones en Centro y Sur América por más de 35 años. Es el investigador principal del proyecto Panamá PIRE (Partnerships in Internal Research and Education) de la Fundación Nacional de Ciencia de Estados Unidos (NSF).

Tigre dientes de sable

Jueves 14 de febrero de 2013
6:30 pm
Parque Explora
Entrada libre
Cupo limitado

Patrocina: Apoyan:

Inset 8.1

give the talk in Spanish (Inset 8.1), which was not easy for me, but I had to do it to reach my audience. I decided not to read a prepared text because talks like these tend to be boring, but rather I would speak from carefully scripted PowerPoint slides. I prepared and practiced my talk more than I had ever done for any presentation in English. One of my graduate students, a Colombian herself, patiently helped me to hone the Spanish. It went well, and my audience prediction seemed largely correct. An added consideration of this process was that I gained a more direct appreciation of what non-native English speakers, such as my international graduate students, have had to deal with during their education in the United States. They do not just have to contend with the fear of public speaking, likely inherent in most cultures, but compounding this was the added dimension of presenting in English when it is not their native language.

> Museums must attract diverse visitors, or risk irrelevance.
>
> Levitt (2015)

The discussion above is primarily focused on bilingual – mixed English- and Spanish-speaking – audiences. What happens, however, if your audience is multilingual, that is consisting of more than two primary languages? Sometimes programming will be done in the dominant second language, or traditionally with exhibit headphones that switch to the dominant languages of the visitor. With the advent of smartphones, programming can be multilingual and customized for individual user experience. An extreme example of a multicultural audience is the New York Hall of Science located in Flushing (Queens; NYHS, 2017). The majority of its visitors are from the local community, which is culturally highly diverse, with more than 50 languages other than English spoken. In this case, having exhibits in only the dominant languages excludes, or presents a barrier to learning for, those groups that are in a minority as visitors. NYHS is by no means in a unique situation and many other informal STEM audiences are now, or in the future will become, multicultural.

Face-to-Face Audiences

Audiences can be reached in many ways, although traditionally the most common mode is face-to-face communication. Face-to-face, sometimes abbreviated F2F, is a time-honored means of communication. Despite almost universal discomfort with the process, public speaking is the *sine qua non* for getting one's point across. F2F can be in structured settings, typically including formal classroom, non-formal, or even informal (impromptu) talks. These audiences allow the presenter the opportunity to actually see their audience as they are presenting. This allows the speaker to "size up" the audience and obtain feedback (e.g., body language) in real time. It also facilitates two-way communication via questions and answers, typically at the end of the talk. Formal public presentations are traditionally less than an hour long; shorter talks in the 10–20-minute range are characteristic of large professional meetings, but also as parts of packed agendas in civic organizations (e.g., Rotary). In order to be successful with F2F communication, the presenter needs to do their homework about the probable composition of the audience. This is by no means a perfect process (as I describe in my lament about the fossil club above), but it will help to optimize the talk. A limitation of the F2F presentation is that the audience has to be physically present; this potentially presents a barrier to accessibility.

Online, Blended, and Social Media Audiences

This category includes a variety of "cyberenabled" audiences that are reached with the help of technology. These can include both F2F and one-way communication. Webinars, or interactive, online lectures and talks, did not exist before the advent of Web 2.0 connectivity. They have become increasingly more widespread and will likely continue to increase in popularity, although the market may become saturated. Nevertheless, the great benefit of web-based presentations is access. As long as the intended participants have access to the internet, then learners can join the webinar in real time, or if recorded, watch the video later. Online access also promotes blended audiences, e.g., a talk presented in real time that is also simulcast to an online audience.

    Another rapidly growing online audience for STEM learning is via social media. It is hard to keep up on social media trends and audience analysis. I've heard disparaging comments that Facebook is for "older people," although this likely is an overgeneralization. For scientists who choose to use social media to promote their profession, Twitter seems to be the most popular method, although many of these social media platforms are ephemeral. The segmentation, strengths and weaknesses, and demographics of the various social media platforms are complex and fluid (Bik & Goldstein, 2013). It remains to be seen what will persist or become dominant in the future. Although the discussion here serves as a brief introduction within the context of audience, more will be presented about the uses of social media for cyber-related Broader Impact activities in Chapter 15.

Written Narrative

When I was preparing the proposal to write this book, my editor asked who my intended audience would be. This was an important first step. It allowed me to both focus on the content and scope of the book, and also predict who might read it. The written word is, of course, a pervasive and lasting means of communicating. Sometimes your audience can be predicted; in other cases, this is more difficult to know. Nevertheless, it is best not to start writing until you know your audience. Likewise, audiences can differ in subtle ways; even within our own science domain, an article written for one scientific journal (e.g., *Journal of Paleontology*) may be pitched differently than if written for another, broader scientific journal (e.g., *Science*). A primary concern about writing articles and oftentimes books (like this one) is that these kinds of communications rarely receive feedback from the reader; they are therefore primarily a unidirectional means of reaching an audience. There are opportunities for book reviews, written rebuttals, letters to the editors, and such, but typically these only represent a small fraction of the author's audience. An online reviewer (Amazon, 2018) of my 1992 *Fossil Horses* book laments that "It fails to dramatise its subject and to attract a 'lay audience.'" This was never the intent of the book, but in an era of open communication, readers are free to express their opinions.

To a certain extent, Web 2.0 has changed the landscape of the written word to one that now provides more opportunities for bidirectional communication between the author and their audience. These can be seen with blogs and social media posts in which comments are freely made and even encouraged by the platform. With the emphasis on learning being bidirectional, or multidirectional, responses and written feedback loops will likely be more common in the future.

Traditional Media (TV, Film, Print, Radio)

Traditional media have many benefits and advantages in terms of reaching potentially interested informal learning audiences – that is, those who can freely choose to learn about science and STEM. More than 60 million people of the general population in the United States are "intellectually curious" about science (Britt, 2006). Similar in size to those in K–12 education, this audience represents a potentially huge opportunity, one that is unrivaled by almost any other outreach strategy. However, certain mainstays of the traditional media market have plummeted in recent decades. This decrease is likely a result of increased content available on the internet. For example,

Box 8.2 The Martian (2015)

Based on the novel with the same name (Weir, 2011), the 2015 movie version of *The Martian* was a hit that grossed $630 million worldwide (Box Office Mojo, 2017). In 2016 it was nominated for seven Academy Awards (although it did not win any of these) and it won the 2016 Golden Globes categories of Best Motion Picture and Best Director (Ridley Scott; IMDb, 2015).

Before I watched *The Martian* in the local movie theater, a botanist colleague of mine said "it is great, you need to see it!" It focuses on the plight of an astronaut (played by Matt Damon) stranded on Mars. Faced with the threat of dying after supplies run out, he remarks: "I will have to science the *heck* [another word used in movie] out of this!" Of most of the recent Hollywood motion pictures intended for the general public market, *The Martian* is exemplary in terms of communicating science and STEM content. Although there are some inconsistencies and misconceptions, in general this movie is considered fairly accurate in terms of its portrayals of both the scientific process and STEM content (Brody, 2015; Zubrin, 2015). It includes science (botany, geology), technology, engineering, and math to solve problems on Mars as well as back at the mission command center in the United States. It demonstrates creativity, flexibility, trial and error, and innovation, all of which are part of the scientific process.

retail sales of magazines in the United States dropped from 103 million in 2014 to about 75 million in 2016. Likewise, revenues from newspaper subscriptions continue to decline, from $33.6 billion in 2011 to $30.5 billion in 2016. The print media industry workforce continues to struggle, with the number of employees falling from 257,800 in 2010 to 183,200 in March 2016 (Statista, 2017). In a similar vein, despite the societal need for authoritative science content presented in an accessible manner (Rehman, 2013), the demand for science writers has fallen, with many newspapers abandoning in-house science news desks. In no segment of the industry has the internet revolution taken such a toll as in print media; its audiences have decreased in favor of online and other forms of accessible media through which they can learn about science.

Science and nature films and documentaries are the mainstay of several series on public TV, and Ira Flatow's weekly *Science Friday* (Science Friday, 2017) is a consistent purveyor of trustworthy and timely content on radio. Feature films dealing with STEM content presented in traditional movie theaters are not universally popular, although there have been blockbusters like *The Martian* (IMDb, 2015; Box 8.2) and *Jurassic Park* (IMDb, 1993), which blend science fiction with STEM. Also in a twenty-first-century context, the rise of online streaming such as through Netflix provides opportunities for access to content in a way that is of interest to the viewing public. With regard to the latter, YouTube is also revolutionizing access, and the influence of this social media platform will likely persist in the future. Thus, the ways in which audiences are accessing science content today is changing dramatically and will likely continue to do so in the future.

Audience Demographics

Along with all of the ways that audiences can be analyzed, a basic understanding of demographics is likewise fundamentally important to effective science communication. The field of demographics is the study and analysis of vital and social statistics and the characteristics of a population and groups that belong to it. Demographics will also become of increasing importance in order to reach underserved, underrepresented, and marginalized audiences (also see Chapter 9).

A set of basic demographic characteristics, including gender (sex), age, and cultural background (race, ethnicity), exist to better understand audiences. In addition, a bewildering array of other demographic characteristics exist that are potentially interesting as well. For example, if museums want to know more about their visitors, relevant demographic data might include: (1) their zip code; (2) whether they came alone or as a family or school group; and (3) primary language spoken. All of these would be of interest in optimizing programming to best serve their visitors. Some demographics, such as marital status and income, are more intrusive. People generally do not like to take surveys, so a good rule of thumb is to only ask those questions that are relevant to the desired information. If you only want to know language spoken, you do not need to know income.

When collecting demographic data, you may need IRB (Institutional Review Board) (e.g., UF IRB, 2017) approval because of the involvement of human subjects. If the intent of collecting data is only for internal planning and evaluation, IRB approval may not be required. However, if the possibility exists to publish or communicate to an external audience (outside your institution), then IRB approval is likely required. On more than one occasion, a student in my lab has decided after the fact that they wanted to present interesting survey results; the process of retroactive IRB approval is painful and not recommended. Approval also includes the requirement of informed consent. If minors are involved, then it is mandatory that parental approval is also received prior to conducting the surveys. This practice is required whether or not the data collected will be used for internal or external communication.

Audience Trends: Case Study of Our Museum Visitation

Since its founding a century ago, our museum – the FLMNH – has had public exhibits and related programs that attract visitors. One metric of visitor analysis is attendance: a museum that has one million visitors per year likely can be argued to have a broader reach than one that has an attendance of 10,000 visitors per year. Thus, in addition to demographics, as described above, visitor attendance can be an important component of audience analysis.

So far as our archives indicate, the earliest attendance data were recorded in 1939 (Fig. 8.4). During that time, our museum was located off-campus within the downtown Gainesville business district. Our exhibits occupied several floors of an office building. For the next few years thereafter, our visitor numbers were estimated to be about 10,000–20,000 per year, although accurate counts were not recorded. In the early 1970s we moved into a new building on the UF campus, Dickinson Hall. Our visitor numbers grew until insufficient parking for public visitors, most of whom were unfamiliar with our campus, became a critical barrier to access. In 1997 the exhibits and public programs function of our museum moved to a new public education building – Powell Hall – which had ample parking, and during the next decade produced a significantly expanded suite of exhibits and public programs. The move to the new exhibit facility also coincided with a more deliberate effort to carefully track visitor attendance, mostly via a "clicker" at the front desk and other estimates of attendance. The apparent dip in attendance after we opened this new facility at Powell Hall in the late 1990s (Fig. 8.4) likely relates to better record keeping, relative to more inaccurate estimates at the previous facility. Starting in 1997 and for the next several years, our annual visitation was accurately determined to be 150,000–200,000. The notable exception during this interval was the spike in 2001 that resulted from the traveling exhibit *T. rex Sue* (90,000 visitors during its three-month period in the venue). In 2004 we opened a new wing of our museum dedicated to Lepidoptera biodiversity, including the Butterfly Rainforest, the latter of which has helped overall attendance (except for a few down years after the Great Recession).

As exemplified in Fig. 8.4, the traditional model of museum visitor numbers is based on physical visitors "in the door" to view exhibits and participate in public programs. It is known, however, that this model only captures a part of

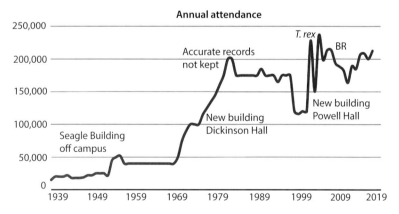

Fig. 8.4 Growth of visitors at the Florida Museum of Natural History, 1939 to present; BR, Butterfly Rainforest (unpublished reports and data, on file in the FLMNH archives).

the overall audience impacted. This is exemplified when museums have out-reach programs, such as during K–12 school visits and in informal settings such as fairs and festivals, and other off-site public events (these will be discussed more in Chapters 12 and 14). The problem with these audience components is that the numbers of visitors have been difficult to accurately record for many of these kinds of events.

Another kind of audience "visits" museums via the internet and social media (Fig. 8.5). This audience has its own characteristics, including the fact that many of these may never actually visit the museum physically. Nevertheless, if one is concerned about broad impact and reach, then those people that connect via social media are important, and also have the added benefit that analytics (e.g., numbers) are readily available. For FLMNH, although about 200,000 people visit our physical museum each year, millions access it via the internet and communicate about it via social media (FLMNH, 2018). If audience reach is important to museums for assessing impact, then the final tally might include: (1) physical visitation to the host museum; (2) visitors at other venues that view traveling exhibitions produced by a museum

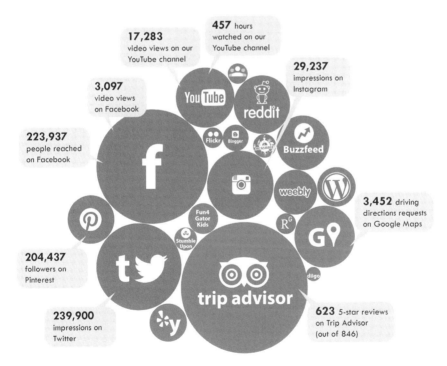

Fig. 8.5 Analysis of a 30-day engagement period during October and November 2018 for the Florida Museum of Natural History via social media "visits." Depending upon how one counts the significance of these numbers and metrics, our social media presence during this time is in the hundreds of thousands (FLMNH, Radha Krueger graphic).

(e.g., such as *Megalodon*; FLMNH, 2018); (3) visitors to museum-sponsored events at public talks, fairs, and festivals; (4) cyberexhibits (e.g., *Fossil Horses in Cyberspace*; FHC, 2017); (5) visits to the museum's website; and (6) some metric that captures the emerging involvement of social media.

Concluding Comments

As can be seen, audiences are complex, frequently overlap in characteristics, and are oftentimes heterogeneous. In order for a scientist to optimize the impact of their communication, it is best practice to study and "know your audience." Yet, even with advance preparation, the actual audience may not be correctly predicted in every instance. Careful planning will, however, minimize times when as a speaker one misses the mark for an effective presentation. Likewise, the landscape of audiences is changing dramatically in the twenty-first century, particularly with the advent of online and social media platforms.

In the next chapter we will learn that in terms of access to science, not all audiences are created equal. For a variety of reasons, it is important for scientists to reach out to a variety of audiences and be mindful of those that are underserved and underrepresented. In so doing, we are exemplifying social responsibility as well as providing equal access and opportunity for all.

9 Diversity, Equity, and Inclusion

Marginalized STEM Learners

In my Broader Impacts class, I present a hypothetical scenario in which the students are program officers of a foundation that provides grants for STEM learning. They are considering two proposals, but only have sufficient funds for one. One proposal seeks to promote STEM learning in maximum-security prisons, where most of the inmates are serving life sentences, with little chance of parole. The other proposal promotes STEM learning for the elderly in nursing homes. All of the proposal reviews are in, and the decision needs to be made to fund one of these equally competitive proposals. Which one should be funded?

There is no right answer, but students engage in debate on both sides. Some of them have had family members in nursing homes and thus feel a connection to this proposal. On the other hand, prisoners potentially have their lives ahead of them and might benefit from social rehabilitation through STEM learning. In the final analysis, therefore, both scenarios could be debated as being the worthy choice for funding. This case study thus explores the paradox of investing in marginalized audiences, such as those in prisons and nursing homes.

Introduction

A fundamental part of the National Science Foundation's (NSF) mission is the commitment to diversity, equity, and inclusion (DEI; Box 9.1). This is not a new concept, but rather stems from the vision of Bush (1945). These core values have been codified in various NSF documents since that time (e.g., NSF, 2007), as well as strategic initiatives such as INCLUDES (NSF, 2017n). Any scientist wanting to submit competitive proposals to NSF is well advised to include an aspect of DEI within the framework of the projects. This component is best scaled for the size of the project; smaller projects can have these elements embedded within the proposal (e.g., recruitment of graduate students), whereas larger projects might benefit from major components and Broader Impacts initiatives involving DEI.

This chapter will cover some of the important themes that pertain to DEI. Related topics are also touched upon, including bias, broadening participation, discrimination, and equality. All of these factors relate to human capital and the progress of STEM. These components do not just potentially enhance the ideas and perspectives brought into a project, they are simply the right thing to do in terms of social responsibility in the twenty-first century.

Box 9.1 Definitions

Bias: A form of prejudice that results from our tendency and needs to classify individuals into categories.*

Broadening participation [also **representation**]: Providing for individuals from underrepresented groups as well as institutions and geographic areas that do not participate at rates comparable to others (NSF, 2008).

Discrimination: The unequal treatment of members of various groups, based on conscious or unconscious prejudice, which favor one group over others on differences of race, gender, economic class, sexual orientation, physical ability, religion, language, age, national identity, religion, and other categories.*

Diversity: Socially, it refers to the wide range of identities. A broad [sic, term that] includes race, ethnicity, gender, age, national origin, religion, disability, sexual orientation, socioeconomic status, education, marital status, language, veteran status, physical appearance, etc. It also involves different ideas, perspectives, and values.*

Equality: The state of being equal, especially in status, rights, or opportunities. (Oxford Living Dictionaries, 2019)

Equity: The fair treatment, access, opportunity and advancement for all people, while at the same time striving to identify and eliminate barriers that have prevented the full participation of some groups. The principle of equity acknowledges that there are historically underserved and underrepresented populations and that fairness regarding these unbalanced conditions is needed to assist in the provision of adequate opportunities to all groups.*

Inclusion: The act of creating environments in which any individual or group can be and feel welcomed, respected, supported and valued as a fully participating member. An inclusive and welcoming climate embraces differences and offers respect in words and actions for all people.*

* Source: University of Washington (2019) verbatim, also an excellent source for many other relevant definitions.

Marginalized Audiences

Numerous segments of society are traditionally treated as a low priority for investments in STEM learning. In this section we follow up the anecdote at the beginning of this chapter, as well as discuss other audiences, including homeless adults.

Prisoners

In 2014, the incarcerated population in the United States was about 2.2 million in state and federal prisons, jails, and other correctional facilities. This number represents about 0.7 percent of the country's population, including about 60,000 juveniles. Of relevance to minority demographics (also see below), a disproportionately large percentage of African-Americans and Hispanics make up prison populations. Studies have shown that, among this general population, educational programs in prisons promote higher rates of successful reentry to society. Within the juvenile population, many of the inmates have not yet graduated from K–12 schools. As such, programs have been developed to address the education of this audience (USBJS, 2017).

Several programs have been developed that promote STEM learning in prisons and youth correctional facilities (e.g., Kunen, 2017; *New York Times*, 2017;). Zaspel (2017) describes a crowdsourcing program in which incarcerated men at the Logansport (Indiana) Youth Correctional Facility transcribe data about aquatic insects in museum collections. Princeton University's Council on Science and Technology (2017) recently sponsored a public event titled "The Prison and the Academy: STEM Education and Prisoner Reentry." Bard College in New York has started an initiative in which over the 16-year span of the program more than 450 incarcerated learners from both men's and women's prisons have received degrees. In addition to this initiative, numerous colleges throughout the United States provide college educations to thousands of prisoners each year. In reference to a specific program funded by the state of New York, it was found that:

> The recidivism rate is 4 percent for inmates who participate in the program and a mere 2 percent for those who earn degrees in prison, compared with about 40 percent for the New York State prison system as a whole. (*New York Times*, 2017)

Biologist Nalini Nadkarni has been a leader in promoting STEM learning to marginalized or non-traditional audiences. She developed the Sustainability in Prisons Project (SPP, 2017), originally started in Washington in 2011 (Fig. 9.1), but more recently in Utah, where she is a professor at the University of Utah. A hallmark of Nadkarni's prison outreach and education is not just getting STEM learning into the prisons, but also pushing active engagement of prisoners, for example in raising endangered plant species.

Fig. 9.1 Nalini Nadkarni working with prisoners at the Stafford Creek Corrections Center, Aberdeen, Washington (Benjamin Drummond and Sara Joy Steele photo).

Other Marginalized Audiences

In addition to prisons, other segments of society are marginalized in terms of access to STEM learning. Little evidence exists in the literature about programs that reach out to residents of nursing homes, or how these programs might benefit society. On a related theme, an innovative program at the Providence Mount St. Vincent assisted living facility in Seattle, Washington, brings 125 preschool children in each week to interact with the elderly. Jansen (2016) discusses the positive intergenerational benefits for both the toddlers and the elderly. This is a model program, and others focusing on the benefits of visits to similar venues have yet to be reported.

Another group – the homeless – are also typically underrepresented in initiatives to promote STEM. Clearly it is difficult to find and engage these kinds of audiences, although they should not be forgotten. Homelessness can affect people of different ages, from those of school age to older adults. While initiatives exist that provide education for K–12 homeless populations (e.g., Homes for the Homeless, 2017), the magnitude of the problem in the United States is sobering. Superville (2017) estimates that about one million K–12 students in the United States are homeless. In the 2015–2016 school year, a staggering 100,000 (10 percent) of students in New York City schools were homeless. In Dallas, Texas, a local nonprofit support group, After8thtoEducate, turns an elementary school with beds into shelters and provides medical services and tutoring (Superville, 2017). Thus, schools can provide a physical place for learning and daily safe haven for these homeless children. In contrast, however, there seems to be a lack of venue and thus opportunity for STEM learning for older homeless individuals. It is not surprising that homeless adults are less engaged in these kinds of learning, given that they likely have more immediate concerns on their minds, such as food and shelter.

Women in STEM

The traditional stereotype of a scientist is an older white male, possibly with a gray beard, lab coat, and long hair (e.g., Brooks, 2012; Leeming, 2017). In many respects STEM also suffers from "good old boy" politics and biases. One would like to think that this culture is changing, but there is still a long way to go. A good example of such inequality is with women in STEM. As also described in Chapter 1, although females make up half of the US population, they only account for about one-quarter of the STEM workforce. This is in contrast to China, where women make up 40 percent of this workforce. The world average is 30 percent (Landivar, 2013; Wu, 2016). Thus, despite its status as a world leader, the United States lags behind in terms of women participating in STEM careers. In addition, women in STEM are underrepresented in leadership positions (Jarvis, 2017). They also lag behind their male peers in terms of compensation, with women being paid significantly less than men (AAUW, 2015; Davidson, 2015; Inset 9.1). Women also face career challenges with regard to families and child-rearing (McNutt, 2013). Despite efforts toward equality, the underrepresented percentage of women in some STEM careers has remained static – there has been little improvement in the twenty-first century (Bidwell, 2015).

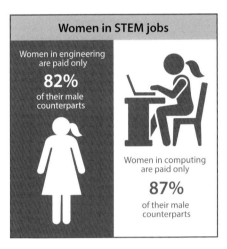

Inset 9.1

The reasons for underrepresentation of women in STEM are complex. Of relevance here is the concept of the "leaky pipeline" (Fig. 1.4), which is a model for how and when females leave the STEM career path. It can occur at an early age from the lack of suitable role models or because of social pressure in which adolescent girls no longer think that being a scientist is "cool." Other factors include pervasive, negative stereotypes about what scientists look like (Leeming, 2017) and how this profession has traditionally been dominated by males.

Coercion and harassment are other aspects of the underrepresentation of women in STEM. Coercion can come from a dominant role of a major professor, supervisor, or more senior authority figures; harassment can take many forms. In recent years there has been a general increase in awareness of the pervasiveness of sexual harassment, mostly of women, at professional meetings. As an example, a colleague would typically counsel her female graduate students about the unprofessional and inappropriate environment, exacerbated by alcohol, during poster sessions at national meetings. Many professional societies in STEM have developed, or are in the process of developing, written codes of conduct and ethics policies that guide behavior not just at national meetings, but in other circumstances, such as in the laboratory and even on sponsored field trips. These practices, however, have typically been slow to become engrained in the cultures of professional societies.

Starting in 2017, the Geological Society of America began a program called RISE (**R**espectful **I**nclusive **S**cientific **E**vents; GSA, 2017). Society leaders and other rank-and-file members can participate in training that heightens awareness and teaches participants how to respond to specific instances of sexual harassment and other forms of inappropriate behavior that might make individuals feel unwelcome. Professional societies take these measures seriously, and consequences for the perpetrator can include expulsion from the meeting or, in particularly egregious circumstances, from the society. It is hoped that these kinds of measures will prevent the loss of women and other underrepresented individuals from the ranks of STEM professionals, although this is just one facet of the overall systemic problem that still exists.

Box 9.2 Hypothetical Case Study: Broadening Representation, Diversity, and Inclusion

You are the president of a major scientific society. At your annual society meeting you are approached on multiple occasions by rank-and-file members who express frustration and concern about the lack of diversity and inclusiveness within the society and that many "good old boy" biases persist. This is exemplified, for example, by the fact that the past three presidents were stereotypical older white males, and most of the prestigious society awards have been made to men, although half of the 1,000 society members are female. You fully agree with these examples as well as your own personal perceptions that this issue needs to be addressed. As president, you are therefore compelled to act. What would you do?

This case study was given to my 2016 Broader Impacts graduate class. They developed the following ideas during a brainstorming session:

Case Study

Evaluate way Presidents are elected

Look at stats for applications for awards

Create advisory panel of women and minorities

Create prestigious award for early
career professionals

Focused symposia on women, diversity,
work-life balance, etc.

Opportunities for groups to meet and talk

Evaluate how many women or minorities are active

Award committees should represent
demographics of the society.

> ## Box 9.2 (cont.)
>
> While this scenario is presented as a hypothetical case, in certain respects it represents one of my societies. Thanks to concerned leadership of that society, this matter is taken seriously. Several tangible actions have been set into motion to improve this situation, including the formation of a Diversity Committee, diversity training, active messaging to the membership, and active recruitment of women and minorities into leadership positions.

Underrepresented Minorities and Underserved Audiences

The US population is becoming increasingly diverse. In addition to women, several groups are not well represented in STEM relative to their proportions in the general population. These underrepresented minorities (URMs) include Hispanics and Latinxs (17.8 percent), blacks or African-Americans (13 percent), and American Indians and Alaskan Natives (1.3 percent; US Census Bureau, 2017b). Although at the beginning of the twenty-first century these groups together represented about 30 percent of the US population, by 2050 they will represent 40 percent. Of these, Hispanics and Latinxs are predicted during this half-century to be the fastest increasing minority group in the United States (NACME, 2017). Despite their percentage representation, these minorities are significantly underrepresented in science and engineering jobs relative to whites (Fig. 9.2). Similar to what is seen for women, the causes of this underrepresentation are complex. A major factor is that for STEM careers a college degree is required, but minority representation, also resulting from a leaky pipeline, decreases as we move up the degree ladder.

Over the past several decades numerous programs have been developed to make the science and engineering workforce more diverse, and to broaden representation so that the percentages of underrepresented minorities reflect those in the general US population. While some strides have been made, as a society we have yet to achieve equality in relative representation (Fig. 9.2). HBCUs (**h**istorically **b**lack **c**olleges and **u**niversities) and HHEs (**h**igh **H**ispanic **e**nrollment institutions; NSF, 2017q [another term used is Hispanic-serving institution, HSI]) are focused on serving these underrepresented minorities. National professional organizations such as the Society for Advancement of Chicanos/Hispanics and Native Americans in Science (SACNAS, 2018) and the National Association of Black Geoscientists (NABG, 2018) seek to promote retention in STEM-related majors and careers for the underrepresented minorities they serve.

In addition to the formal recognition of URMs, other demographics represent underserved audiences. Sometimes the terms "underserved" and "underrepresented" are used interchangeably or have very similar practical meanings. For example, urban inner-city populations frequently consist of underrepresented

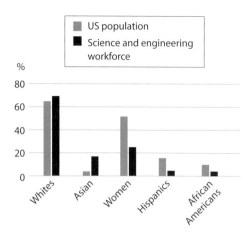

Fig. 9.2 Percentages of demographic groups (age 21 or older) in the US population versus their respective percentage in the science and engineering workforce (NSB, 2016).

minorities, particularly Hispanics/Latinxs and African-Americans. This is, however, not always the case; for example, Asian-Americans frequently represent significant percentages of urban inner-city populations, but they are not underrepresented minorities in terms of their participation in STEM (Fig. 9.2).

Irrespective of their racial and ethnic composition, underserved audiences typically have barriers to access. A classic example of this is rural underserved populations. Although these are classically white elsewhere in the United States, large enclaves of African-Americans exist in the rural south. In addition, over the past several decades, the percentage of Hispanics and Latinxs in rural areas has increased in many parts of the United States (Effland & Kassel, no date; PRB, 2017). In this case, and like in urban settings, the underrepresented (Hispanic/Latinx) minority is also part of the rural underserved demographic. Generally, rural under-served communities, whether they are underrepresented minorities or whites, suffer from the lack of access to cultural institutions, such as museums, that are typically available in more populated regions. In contrast, and on the positive side, rural underserved communities benefit from being close to natural settings, unlike the pervasive problem in heavily urbanized cities lacking parks and open spaces (Cox et al., 2017).

Strategies exist that can mitigate the lack of access. Displays at rural fairs, and "pop-up" museums (Fig. 9.3), can bring science into the community rather than expecting these underserved audiences to come to museums. Delivering science content using other modes, such as TV, radio, the internet, and via mobile phone technology (including apps), are other equalizers that can potentially broaden representation of underserved audiences (also see Chapter 12). The recent explosion of 3D-printing technology can also broaden access and participation of rural under-served communities. For example, the iDigFossils project has brought 3D-printed fossils into rural underserved communities in New Mexico via local libraries (iDigFossils, 2017; Fig. 9.4).

Disabilities and Barriers

> true commitment to the inclusion of people with disabilities remains an exceptional practice – not the norm.
>
> (Reich, 2012: 24)

Kinds of disabilities
Autism
Chronic illness
Hearing loss and deafness
Intellectual disability
Learning disability
Memory loss
Mental health
Physical disability
Speech and language disorders
Vision loss and blindness

Inset 9.2

In 2010 nearly one in five (19 percent) of people reported that they have a disability, with half of these being severe (US Census Bureau, 2012). Disabilities can cut across socioeconomic and demographic spectra and take a variety of forms (Inset 9.2; LDS, 2017). They can also limit employment opportunities in the US science and engineering workforce (Fig. 9.5; NSF 2017u). With the aging of our society and recognition of the kinds of disabilities that were previously not categorized, the representation of persons with disabilities within the US population will likely increase in the future.

The Americans with Disabilities Act (ADA) was passed in 1990 (EEOC, 2017) to protect people with disabilities from discrimination and to provide them with equal access and opportunities. Similar to some other social programs enacted at the federal level, the ADA has met with limited success. While the ADA has increased public awareness of persons with disabilities, it is a voluntary program in some sectors (e.g., without federal funding) and it has therefore been inconsistently implemented (Cardillo, 2015).

Because of federal funding to our institution, when we built our new exhibits museum in the late 1990s we were aware of the importance of ADA compliance. While we knew that some of our visitors would have some range of disabilities (Fig. 9.6), we were not aware of the frequency and range of these. Obvious accommodations for visitors with physical and mobility concerns included ADA-accessible architectural design of the building and its exhibits (e.g., handicap access and no stairs). We also understood that exhibit videos needed to be closed captioned, and that certain exhibit components could be touched for the sight-challenged. We were, however, less attuned to other disabilities and this affected our ability to be more responsive to them. In recent years we have developed focused programs for the special needs of our visitors. For example, we host groups with intellectually disabled visitors and have piloted early morning visits for autistic visitors in a quieter environment, as well as lowered light levels (Rais, 2012).

Fig. 9.3 Fish pop-up graphic at a local fish festival in Cedar Key, Florida (FLMNH graphic).

Fig. 9.4 Saturday morning STEM activity with 3D fossils at the El Rito Library, New Mexico (November, 2017). UF paleontology graduate student Sean Moran (upper right) is explaining the arrangement of 3D-printed teeth of the giant extinct shark *Megalodon* on the display in the background (author photo).

Box 9.3 Geerat Vermeij: An Inspiration in Paleontology

(Geerat Vermeij permission)

Love your subject, be prepared to work hard, don't be discouraged by the occasional failure, be willing to take risks, get as much basic science and mathematics as you can take and perhaps above all display a reasoned self confidence without carrying a chip on your shoulder.

(Vermeij, in NCBYS, 2017)

Dr. Geerat Vermeij, Distinguished Professor at the University of California – Davis, is an expert on fossil mollusks and evolution interpreted from the fossil record. He received his PhD in an astoundingly short time – three years – from Yale University. He has had a successful career and received many well-deserved accolades, including the MacArthur Fellowship "Genius Award" in 1992, and the Paleontological Society Medal in 2006. Geerat's scholarly publications include many peer-reviewed articles and books that integrate a broad range of subjects, including paleontology, the evolutionary "arms race" among species, and even economics. His autobiography, *Privileged Hands: A Scientific Life*, was first published in 1996 (Vermeij, 1996).

Box 9.3 (cont.)

In addition to his academic pursuits and great accomplishments, Geerat gives back to society, for example as a role model for the National Center for Blind Youth in Science (NCBYS, 2017). This is perhaps not surprising because Geerat lost his sight at the age of three, but this has not slowed him down. On not being able to see, he noted that: "I am a strong disbeliever in seeing things from the point of view of being handicapped, gender, race and all the rest of it" (Yoon, 1995).

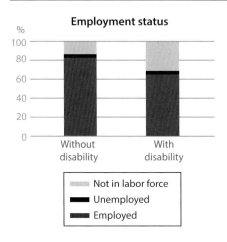

Fig. 9.5 Employment status (percentage) in US science and engineering for persons without and with disabilities in 2015 (NSF, 2017q).

Other Barriers

The percentage of disabilities within the population increases with age (US Census Bureau, 2017b), and the same is true for our museum visitors. Because our visitor demographics indicate a considerable age range of lifelong learners, we know that other barriers exist that likely affect their experience. Signage can be a barrier to an optimal visitor experience, not just in wayfinding, but also in the interpretive parts of exhibits. We follow ADA guidelines in terms of wayfinding signage, and the text (font size) in our exhibits follow national guidelines and best practices (e.g., Serrell, 2015). Although not strictly related to disability, barriers exist for other segments of society as well. One of the most glaring examples of this kind of barrier occurs in K–12 education in which students struggling to learn STEM are hampered by language barriers. ESL (**E**nglish as a **s**econd **l**anguage) or ELL (English language learners) programs are therefore more commonplace, particularly in regions with high percentages of, and rapidly growing, minority populations.

Proposers to NSF should understand that part of broadening participation also means that in addition to women and minorities, inclusion of disabled participants can enhance a Broader Impacts plan. Implementing these in innovative ways to disabled audiences will almost certainly enhance the overall quality of the visitor experience.

Fig. 9.6 Persons with disabilities visiting museums. Top: Dinosaur hall at the American Museum of Natural History in 1929 (AMNH photo). Bottom: Florida Museum of Natural History (FLMNH, Jeff Gage photo).

DEI Programs at NSF

The National Science Foundation takes DEI seriously, not just within the institutional culture of the organization, but also in the programs that it funds. With regard to gender, the NSF has programs such as ADVANCE (NSF, 2017r), which are targeted at promoting women in STEM. Many of NSF's directorates have their own DEI initiatives, such as that of GEO (NSF, 2017s). Although it is not as well known among proposers, an entire division – Human Resource Development (HRD) – within the EHR is focused on programs in which DEI are a primary motivating factor for funding. These programs can be agency-wide (e.g., ADVANCE) or specific solicitations from HRD (e.g., for Latinx- or Hispanic-serving institutions). More recently, as part of the agenda of the current NSF director, France Córdova, DEI is promoted with the INCLUDES program (NSF, 2017n).

When developing a Broader Impacts plan, an appropriate strategy with each of these targeted programs is to identify an underrepresented audience and embed the proposed research and educational activities within a framework of reaching this audience. In these cases, evaluation and demonstration of positive outcomes are as important as any research that is proposed. Thus, the culture of an HRD project typically has a different focus from a standard research proposal. In addition to the scientific research, for projects that have a significant DEI component, it is a good idea to involve a social scientist or science educator with a background in learning research or educational psychology on the project. In so doing, evaluation of the DEI component of the project will receive the attention that it deserves. It also will likely be received in a more positive light by reviewers of the proposal.

Concluding Comments

While some researchers tend to focus on other aspects of Broader Impacts activities, the 2007 Dear Colleague Letter (NSF, 2007; Inset 2.2) makes it very clear that broadening participation of underrepresented groups is a priority. In the past it may have been sufficient to propose that an effort would be made to recruit minority students within a lab group, but expectations are greater today. For a DEI component of Broader Impacts to be considered innovative, a more inclusive plan is required, one that reaches outside the comfort zone of the ivory tower into underserved and marginalized target audiences, like those described above. In so doing, the scientist is not just aligning with NSF's Broader Impacts expectation, but will also be demonstrating a commitment to social responsibility.

10 Mentoring and Role Models

Why am I a Paleontologist?

This could be answered in many ways, but here I want to focus on 1964–1965, when I was a tenth-grade student at Port Chester High School (New York). In my earth science class our teacher, Gerald Greenstein (Fig. 10.1), assigned each of us to write a term paper. I chose to do mine on the fossil bird *Archaeopteryx*. I spent Saturdays in the Yonkers (New York) public library reading books by paleontologists such as G. G. Simpson, one of the leading scientists of the twentieth century. Mr. Greenstein liked my term paper and encouraged my interest; I also found the course content and his teaching style interesting. This was a formative experience in my career thanks to a teacher.

When I was writing this book, I decided to reach out to Mr. Greenstein and thank him for his impact on me. After more than a half-century I realized that it was a long shot. From school records and after several hours on the internet, I learned that Mr. Greenstein left PCHS in 1968, received his doctorate, remained a K–12 educator, and to my surprise had retired in south Florida. I wrote him a letter thanking him for the influence he had on my life. His wife, Avery, called me and broke the sad news that Dr. Greenstein had passed away in 2012. Avery said that she and their daughter were moved by my letter, as Jerry would have been. Avery asked me to explain the impact that Jerry had on me. I explained that he encouraged my interest in science and that by doing the research for the term paper I found I wanted to pursue a career in paleontology. The latter experience was the first time that I "researched" anything scientific in a formal way. It was fascinating to dig deeper to learn about things, and I also found that I enjoyed writing.

Mentors and Mentoring

Nowadays in academia mentoring is taken seriously and oftentimes is highly structured. This can include many steps along the pipeline, including at-risk

Fig. 10.1 Gerald Greenstein (Peningian yearbook, 1967, retouched).

students transitioning from high school, undergraduates, graduate students, postdocs, and early-career faculty working toward tenure. In reality, mentoring of one form or another, whether it is structured or informal, occurs throughout one's academic career. Published studies from a variety of disciplines, ranging from STEM, to medicine (e.g., Detsky & Baerlocher, 2007), to the humanities (Pye et al., 2016), have highlighted the positive benefits of mentoring, including increased productivity, professional success, and career satisfaction. While mentoring or coaching has been practiced for millennia in academia, over the past several decades it has become more intentional. The National Academy of Sciences (1997) published an important, foundational report that provided recommendations about academic mentoring, titled *Advisor, Teacher, Role Model, Friend: On Being a Mentor to Students in Science and Engineering*.

The traditional notion of the mentor being the expert and the mentee being the novice (or protégé; e.g., Clifford et al., 2014), oftentimes in a supervisor–supervisee relationship, is still commonplace, but other forms of mentoring have become part of this process. These can include mentoring by persons who are not in a supervisory role (thus mitigating the negative effects of power, coercion, and influence), to small-group and peer mentoring, to online communities of mentoring (e.g., Pye et al., 2016). In a modern context, the extent to which mentors should be friends, or surrogate parents, is fraught with mixed messages and potential professional and personal boundary issues. In a recent editorial in *Science* titled "I'm not your mother," Larisa DeSantis, a faculty member at Vanderbilt University, opines:

When I was a graduate student, my advisor made it clear that he was not my father. I got three hugs: when he attended my wedding, when I passed my qualifying exam, and when I had my first child a few months before graduating … I am expected to be nurturing … regardless of the circumstances. But I can only help them [her students] reach their full potential as their mentor – not their mother.

(DeSantis, 2017)

Thus, another consideration for the mentoring process is that societal expectations can differ between males and females, the latter of whom, if they have children, are more likely to have the responsibility of child-rearing as well as being early-career women academics.

In addition to the impact that mentoring has on the next generation, studies have also demonstrated the positive influence that role models can have, particularly those who buck the stereotype of a scientist being an old white guy with scruffy hair and a lab coat. Likewise, institutions such as the National Science Foundation (NSF) have realized that further emphasis must be placed, and attention paid, at certain stages of the career pipeline, including women, postdocs, and underrepresented minorities in science, all of whom are potentially at greater risk of encountering barriers to success and advancement.

In this chapter, I reflect on my experiences as both a mentor and mentee and how, recently, my students have served as near-peer role models in K–12 schools with high numbers of underserved minorities. I then review NSF's targeted mentoring of at-risk segments of the workforce, including postdoctoral fellows and the INCLUDES (NSF, 2017n) program, the latter of which is focused on minorities who are underrepresented in STEM. Most evaluations of mentoring programs primarily focus on the positive benefits for the mentee. We have been working with teachers on NSF-funded projects over the past decade. Results presented below of a summative evaluation of one of our projects provides evidence, not generally understood in academia, of the significant benefits accrued back to the mentors of K–12 teachers.

On Being Mentored within the Ivory Tower and the Real World

Whether it was in a formal role or just happened, all of us have benefitted from being mentored in one way or another. It is inevitable and also has contributed to each of our individual successes in what we do. While we typically think of mentoring within an academic framework, there also can be instances of mentoring outside, in the real world.

In high school I worked in a European-style kitchen at a country club in New Rochelle, New York. For many years, first during the summers and then "moonlighting" as a graduate student, I was mentored by the head chef, Wilhelm

Fig. 10.2 Malcom C. McKenna in northern New Mexico, summer 1972. The field work that we did on this trip provided the spark for my doctoral dissertation (MacFadden, 1976) (author photo).

Niederdorfer, someone for whom I had great respect. Wilhelm was from Austria and he and his family suffered mightily during World War II. From Wilhelm, I learned to take pride in what you do, to work hard, to stay ahead of things, and don't complain (because "you got it pretty good"). I have found these core ideals to be transferable to any profession, including paleontology. Wilhelm and Mr. Greenstein (in this chapter's opening anecdote) mentored me at a formative time in my life.

As an undergraduate, first at the University of Maryland, I was encouraged by two inspirational geology professors, Henry Siegrist and Peter Stifel. After I transferred to Cornell University, I was again mentored by two geology professors, John Wells and Art Bloom. I did an Honors thesis on fossil fishes from New York with Dr. Wells, which helped to further develop my love of research. Dr. Bloom, with whom I took two courses, encouraged me to apply to "the best graduate schools" at a time when I lacked the confidence to do so.

I returned to Cornell on sabbatical almost two decades later. Dr. Wells had passed away, but Dr. Bloom, although retired, was still active. At that time I went to Cornell to begin my book on fossil horses (MacFadden, 1992), something that I found difficult to do because of the daunting scale of the project. Dr. Bloom was a successful book writer, and he one day advised me (paraphrased): "Just start writing it. Set yourself a goal of one page per day. After a year you will have the bulk of your book written and it will be done in a year and a half!" This guidance helped me finish my first book and I have followed this advice ever since. The same advice also seems helpful when I mentor graduate students who struggle to start writing their theses or dissertations.

As a PhD student at Columbia University I was co-mentored by my principal advisor, Malcolm McKenna (Fig. 10.2), and co-advisor, Neil Opdyke, both of whom

Fig. 10.3 Left: Morris and Marie Skinner, *c.*1955 (Rayleigh Emry photo). Right: author and Morris Skinner at the Society of Vertebrate Paleontology meeting in Dallas, Texas, 1973 (author's collection, photographer unknown).

were giants in their fields (Neil was later elected to the National Academy of Sciences). Their mentoring did not come in the form of intentional lab meetings like we see today, but more leading by example. As a graduate student at Columbia I learned from Malcom how to work independently. From Neil I learned the importance of asking interesting and "big" questions in science. Being a graduate student in a challenging PhD program is a heady yet demanding experience. I was influenced by many others, including my fellow graduate students, many of whom knew much more about science than I did.

My graduate education would not have been complete without the mentoring by Morris and Marie Skinner (Fig. 10.3), neither of whom were on my committee nor formally recognized as contributing to my professional development. Morris was a fossil collector and then curator at the American Museum of Natural History in New York (AMNH, where most Columbia University graduate students study paleontology). He was at that time a world authority on fossil horses, and freely encouraged me to develop an active research interest in this fascinating group (MacFadden, 2007). All of my mentors so far were older, white males. I doubt that this was intentional on my part, it just happened that way. I am therefore grateful for the impact that Morris' wife, Marie Skinner, had on my career. A volunteer at the AMNH when I knew her, Marie was a kindhearted, down-to-earth, and nurturing person who helped me deal with the rigors of being a graduate student in a way that none of my official mentors had done; she was always available, and we had long talks about science and life in general.

It is generally thought that most mentoring occurs during one's formal education, and then you are finished. This could not be further from the truth. Mentoring is a continuous process that occurs throughout one's academic career. I have been mentored by many colleagues and collaborators since receiving my PhD, and I suspect this has been the same for most academics. Likewise, I frequently keep in contact with my former students, some of whom still seek advice and guidance long after they have graduated.

Mentoring Style

Just like when we were thrust into college teaching with little or no formal pedagogical training, so too did we become mentors of graduate and under-graduate students with no formal training. As a PhD student, my advisor did not have formal lab meetings or regularly scheduled appointments. Nor did we ever discuss the intentionality of mentoring; it just happened. My advisor led by example by virtue of his hard work and creative intellect. He also typically was available in his office for one-on-one discussions. Over the past 40-plus years as a mentor, I have copied this mentoring style and primarily been available via individual tutoring sessions. These are typically scheduled by email, or if my office door is open it means I am available. Until recently this worked well, but more recently, with seven graduate students, this style became impractical. It made sense for us to have weekly "lab" meetings in addition to individual mentoring, the latter of which I still try to encourage with my students. I also maintain ongoing communication via an online discussion group. Some other lab meetings are focused on reading articles – like a journal club – but we typically do not do this; ours tend to be open discussions of announcements, our research progress, upcoming activities, and plans. I have come to enjoy the weekly lab meetings. The students have reflected on how the lab meeting is a safe place where they can talk without fear of being judged. It is also described as an inclusive environment for those who are less well versed in a particular subject. The sense of belonging to the group, opportunities for networking, and camaraderie are also important.

Mentoring boundaries have typically worked for me. I do not go out drinking with my students or typically socialize with the lab group. As DeSantis (2017) also notes, I'm not buddies with my students. With regard to discussion of personal problems or gossip, these kinds of interactions are not encouraged unless a student feels the need to talk with me about what is happening. I like to keep the professional boundaries, but also realize that one size does not fit all and sometimes students have a need for "fatherly" or friendly advice, but this is not the norm.

Coaching and encouragement are important, particularly to build students' confidence at this critical time in their professional development. Nevertheless, I hope that my students do not think of me as "spoon-feeding" or micromanaging their research and educational program. One of the most important things I learned

> **Quality mentoring**
>
> • Individualized support
> • Academic excellence
> • Humility and empathy
> • Collaboration and networking
> • Impartial advice
> • Respect and trust
> • Honesty and open communication
> • Patience and freedom to explore

Inset 10.1

as a graduate student was how to work independently, with active mentoring only when I was stuck or otherwise needed some direction. This serves us well because after the PhD we are on our own. Working independently is a skill that promotes success.

Sills (2018) introduced a forum in which young scientists from around the world were asked to describe one quality of a mentor that the mentee would try to emulate when they became a mentor themselves. The results are not surprising (Inset 10.1) and they show a range of observations and traits, both academic and personal. It is very likely that many of us experienced good qualities in our mentors that we have tended to copy in our roles as advisors. Not surprisingly, we also tend to avoid other mentoring traits and behaviors of advisors that we did not like during our early careers.

Unanticipated Benefits for K–12 Outreach Mentors

> partnership experiences with teachers can have profound impacts on scientists themselves.
>
> Tanner (2000: 3)

Over the past decade I have collaborated with K–12 teachers, mostly from California and Florida (also see Chapter 11). While much is known about the positive benefits for the mentee (e.g., NAS, 1997), little has been reported about the positive benefits accrued back to mentors. I have come to realize that my mentoring experiences in K–12 settings had positive benefits for both the scientists (mentors) and teachers (mentees) involved. Most scientists considering K–12 outreach as part of their Broader Impacts plan likely do not realize this potentially career-changing benefit for them.

Tanner (2000) reports a study of the impact on the scientist in scientist–teacher partnerships. Her study represents one of only a few that discuss the impact on the mentor. Tanner (2000: 3) describes anecdotally that many scientists understand the value of mentoring teachers, and their comments include that "it's fun" and "it's a chance to get out of the ivory tower" to more self-serving reasons such as "it's an opportunity to develop a scientifically literate public" and "it's a chance to develop advocates for research and research funding." In addition, however, Tanner did formal research in which she used a mixed-methods (qualitative and quantitative) study with classroom observations, interviews, surveys, and pre- and post-assessments. The results of

Tanner's (2000) study indicate that scientist–teacher partnerships can have a profound effect on the scientist-mentors. These include benefits to the scientists as professionals, future educators (the scientist-students), and as individuals, and thus enhance their collaboration, communication, pedagogical, and professional skills (also see SEPAL, 2018). Other studies and descriptions in the literature (e.g., Institute for Clinical Research Education, 2018), although few, likewise describe the positive benefits accrued to the scientist-mentor, including enhanced professional leadership and collaborative skills as well as a sense of satisfaction of contributing to the next generation's professional development.

Similar to these findings, our K–12 outreach has shown us that most scientists, including PhD researchers, graduate students, and undergraduates, enjoy interactions with the teachers. However, we had no formal data to address this notion. We therefore hired a professional evaluator to conduct a summative study of one of our teacher projects. Through validated surveys, structured interviews, and focus groups of scientist-mentors on the Great American Biotic Interchange (GABI) Research Experiences for Teachers (RET) project (GABI RET, 2017), we sought to determine in what ways working with teachers affected the scientists, including their professional practice. Related to this goal, we also developed a list of expected outcomes of our partnership (Davey, 2017; Inset 10.2), from the point of view of the scientist-mentor.

Our scientist-mentor evaluation involved 15 participants, including professors, museum scientists, graduate students, and undergraduate interns. Most had little prior experience working directly with K–12 teachers or in their classrooms. Classroom ("role model") visits after the field paleontological experience became an important part of the professional development, for both teachers and scientists. As a result of the GABI RET project interactions, the scientist-mentors had a mean of 10 collaborations with specific teachers and in the K–12 classroom, although one reported more than 50. Related to expected outcomes (Inset 10.2), the evaluation results (Davey, 2017) indicated significant gains in knowledge and appreciation of K–12 teaching as a profession, lesson planning, STEM learning standards (e.g., NGSS, 2013), and classroom management. With regard to the scientist-mentor practice, the experience resulted in improved teaching and communication skills. The scientist-mentors also greatly enjoyed seeing the teachers' excitement for learning about science, passion for, and dedication to, their profession, and how they translated what they learned in the classroom.

Expected scientist outcomes

Increased
- research productivity
- ability to communicate research
- awareness of K–12 education
- awareness of differences in K–12 and higher education cultures
- enthusiasm for research

Expanded
- outreach
- scientist–teacher community

Improved mentoring skills

Inset 10.2

Pre- and post-assessment of expected outcomes resulted in significant gains ($p \leq 0.01$) in perceived benefits for the scientist-mentors (Fig. 10.4). Comments from our validated survey also provide qualitative evidence of benefits accrued

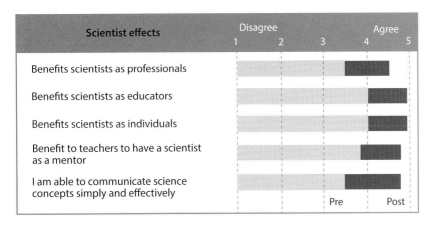

Fig. 10.4 Pre- (light gray, baseline) and post- (dark gray, gain) effects of scientist–teacher partnership on the scientists. All post-intervention gains are significant ($p \leq 0.01$; Davey, 2017).

Box 10.1 Verbatim Open-Ended Responses from the Scientist-Mentors from the GABI RET Summative Evaluation (Davey, 2017: 59–65)

- I think that working with teachers has helped me learn how to communicate science to people in a more clear and concise way.
- The main influence in working with the teachers is that I now think about how my scientific research might have a broader impact on the public.
- I changed my scientific writing style to accommodate a wider audience and to hopefully be a tool to teach science writing and science concepts in the classroom.
- I no longer consider a publication in a peer-reviewed journal as the end goal for a research project. Now I see it as a classroom lesson.
- I now not only feel more comfortable sharing my own research with broader audiences of all ages, but also see the necessity of doing such.
- I was surprised by how patient they [teachers] all are and their perseverance.
- Mentoring is more collaboration than supervisor/employee type relationship.
- I learned that the teachers are highly motivated to learn as much as possible, and to pass along this new information to their students.
- I learned how much teachers sacrifice for their students, and I was surprised to find this is true across teachers from all over the country.
- I became far more interested in science education than I was before and led to my decision to pursue a PhD to merge scientific research with science education/outreach.

back to the scientist-mentors, including: (1) the development of better communications skills; (2) insights into K–12 teachers and their profession; and (3) a common theme that these experiences resulted in rewarding outreach. It is also interesting that most scientist-mentors considered the teachers as peer collaborators instead of an expert–novice ("supervisor–employee") model.

Summary and Recommendations

This one case study of a scientist–teacher partnership (Davey, 2017; GABI RET, 2017) demonstrates the mutual benefits of mentoring. It clearly had a positive and significant impact on the scientist-mentor, and although the results are not reported here, this is also the case for the teachers. The large majority of scientist-mentors responded that they would like to continue to work with teachers in the future. This is likely a common thread or theme that would emerge from other scientists who have worked with STEM teachers, yet it is likely not fully understood or appreciated by the general scientific community. This is unfortunate, not just for their own professional development and sense of satisfaction, but as we will see in the next chapter, collaborating with K–12 educators also provides immense opportunity for Broader Impacts along with societal benefit.

NSF and Mentoring

The National Science Foundation realizes that effective mentoring is important not only for positive outcomes of funded research, but also for professional development of human resources in STEM, particularly for broadening participation of the next generation. Deliberate mentoring plans and strategies are particularly important and typically more explicit within the Education and Human Resources (EHR) Directorate. Nevertheless, these should also be carefully developed in proposals submitted to the STEM research directorates, particularly for larger projects with embedded professional development components. In the past, mentoring was somewhat informal, or less well envisioned, in many NSF proposals. However, during the past decade this practice has become more structured and explicit, particularly with respect to the development of early-career professionals and underrepresented minorities in STEM.

Postdoctoral Fellows

The NSF discourages the use of boilerplate language for the postdoctoral researcher mentoring plan. Each mentoring plan should be tailored to meet the needs of the particular research program and discipline.

UIGC (2018)

The position of a postdoctoral research fellow ("postdocs") in academia exists in the never-never-land between finishing the PhD and the traditional goal of obtaining the first tenure-track job. There are at least two problems with this scenario: (1) tenure-track jobs in academia are becoming more elusive (see, e.g., Fig. 1.5; Blank et al., 2017; Lohr, 2017); and (2) of direct relevance to this discussion, postdocs have traditionally been "unregulated." The latter also means that postdocs sometimes fall between the cracks with regard to the mentoring that they receive. All of these considerations are exacerbated by the hypercompetitive job market, one that in certain disciplines like paleontology remains a tremendous challenge. In many cases, students receiving their PhDs in low-demand STEM fields will have a succession of postdocs (or temporary instructor positions) before they become competitive for a tenure-track position. Thus, the postdoc interval in terms of a young scientist's professional development is critical to their preparation to enter the STEM workforce. It represents a critical transition after the PhD (Costello, 2018).

In response to a growing awareness of the inconsistent mentoring of postdoctoral research fellows, and also stipulations within the America Competes Act (2010), a decade ago NSF (2009d) instituted the requirement of a postdoctoral mentoring plan. I was a program officer at NSF (2009–2010) when this requirement was being instituted. Similar to other new requirements that have been foisted upon uninformed NSF proposers, the early postdoctoral mentoring plans were of inconsistent quality and specificity. At NSF we held internal meetings to evaluate the postdoctoral mentoring plans in proposals submitted to our program. Nevertheless, most of these passed through the compliance screening process, even if they were rudimentary.

As the postdoctoral mentoring requirement became engrained in the framework of proposals submitted to NSF, these gradually improved. Many resources are available to postdocs and mentors to develop a meaningful plan, many of which have been developed by graduate colleges and sponsored research offices at universities (e.g., University of Illinois [UIGC, 2018]). Other organizations, such as the National Postdoctoral Association (2017) also provide valuable resources. Effective postdoc mentoring includes an explicit, written plan and process of professional development (similar in general concept to an individual development plan or IDP (e.g., Clifford et al., 2014; UIGC, 2018). Best practices also include setting defined goals, describing the mode and frequency of communication between postdoc and mentor, specific career-building activities and additional skills development, and methods for periodic performance assessment. Although it is not frequently included in the plan, an explicit description of authorship expectations (e.g., Berg, 2018a; Strange 2008;) and intellectual property will obviate uncomfortable situations during the postdoc's appointment in the mentor's lab.

The collections and research divisions within our museum have two dozen postdocs working within many different research groups. The postdocs have organized a monthly lunch at which time they talk about common issues. In 2017, we informally polled the postdocs and found numerous common themes united them, including: (1) concerns about their future careers; (2) a need for more visibility and emphasis within the museum; and (3) a desire for additional skills development (e.g., in our exhibits museum and public programs). Not surprisingly, there was also inconsistent satisfaction with the mentoring they received. Some of these postdocs are also members of a campus-wide postdoctoral association. Opportunity certainly exists for improving the postdoctoral experience, and this is likely not unique to our unit.

Reviewers of NSF proposals are specifically asked to comment on the postdoctoral mentoring plan. In reality, however, these plans are rarely scrutinized to the extent that their quality would affect the final judgment of the proposal. Thus, at the present time, it is more of a compliance issue. In terms of outcomes, however, a 2003 Sigma Xi survey (Davis, 2005) indicated that a postdoctoral mentoring plan co-created with the mentor early during the postdoctoral appointment was associated with greater success of the postdoc. If we are committed to mentoring the next generation of STEM, then postdoctoral mentoring plans are not just another NSF requirement, but the right thing to do for early-career scientists entering the workforce.

Role Models and Broadening Participation

> Ask children in second grade and upwards to draw a scientist, and you are presented with a white male wearing a white lab coat, glasses and an excess of facial hair.
>
> Brooks (2012)

Five decades ago, the administration of our museum provided us with white lab coats, presumably to make us look more like scientists. Although since that time most of us have not worn lab coats, some of us have otherwise grown into the classic stereotype of what a scientist looks like. There is nothing that I can do about this, except perhaps shave off my beard. Stereotypes are not good, particularly for the next generation aspiring to go into STEM careers. A large body of literature exists on the negative aspects of stereotypes (e.g., Brooks, 2012; Riben et al., 2014), while at the same time underscoring the importance of identity, making STEM inclusive, and providing opportunities for a diverse workforce in the twenty-first century. Role models and effective mentors can be fundamentally important in moving in the right direction with regard to broadening participation.

We have known about the limiting impact of stereotypes for a long time. As you may recall from Chapter 2, in 1945 Vannevar Bush, who was the principal architect of NSF, cited the importance of science and technology advancing through the contributions of different kinds of people. This theme, therefore, has been part of NSF's culture since its beginning. It is further underscored and affirmed via initiatives such as INCLUDES (NSF, 2017n). Despite the general recognition of the importance of a diverse and inclusive STEM workforce, it is still in most cases dominated by stereotypes, and women and minorities mostly remain under-represented, despite efforts to change this inequity. In addition to effective mentoring, another way to improve this situation is encouraging STEM role models that more closely resemble the general population of the United States, with particular emphasis on underrepresented groups.

Role Models and School Visits

Over the past decade my graduate students have been invited by teachers to make presentations to schools in places such as Santa Cruz County, California. These presentations are not about just their science (what they do), but also their journey to become a scientist (who they are). These activities have thus become a student-driven initiative embedded within a long-term project to be involved with K–12 teachers in our research. From the beginning, my graduate students have been an integral part of our K–12 outreach and collaborations. Several of my students have been young Latinas, and the teachers felt that they would make excellent role models for the classroom, particularly in areas with large Hispanic and Latinx student populations.

One of the California teachers, Jill Madden, invited my then-PhD student Catalina Pimiento (Fig. 10.5) to present at Cesar Chavez Middle School, where Jill worked as a middle school earth and environmental teacher. With largely (>90 percent) Latinx students at Cesar Chavez, Jill believed that Catalina would be an excellent role model for young, aspiring students who might not understand the opportunities in STEM – like paleontology. Catalina's visit to Cesar Chavez was anticipated by both Jill and the students, the latter of whom had to draw places via a lottery to get a seat in the classroom where Catalina would talk. By all accounts, Catalina's visit was an inspiration to all, including herself, for the warm and enthusiastic welcome she received. Catalina felt like a "paleo-rock star."

On another visit, to Watsonville High School in southern Santa Cruz County, close to Jill's school, biology teacher Dan Johnston invited Catalina to present to his biology classes. Again, Catalina was a hit with the largely Latinx students. She discussed her journey but also encouraged the students interested in STEM careers and college. In subsequent correspondence with Dan, he further validated the impact of Catalina's visit when he shared with us that, while reading drafts of college applications, he found that one of the students' motivation to pursue STEM was catalyzed by Catalina's visit and individual encouragement. As Catalina was finishing her PhD, and we had time to reflect on her education, she confided in me that her role model trips to schools, particularly with Latinx students (Catalina is from Colombia), were among the highlights of her graduate career.

Many other UF graduate students have done role model visits and presented about science and careers in K–12 schools. This program has organically evolved within my lab group as a result of encouragement from teachers and interest from the

Fig. 10.5 Top: Former UF graduate student Catalina Pimiento during a role model visit to schools in Santa Cruz County, California, c. 2013 (Robert Hoffman photo). Bottom: UF graduate student Michael Ziegler presenting to third-graders in a school in New Orleans, Louisiana (Elizabeth Lewis photo).

UF students. Since Catalina's visits to California, we have done similar school visits in elementary through high schools in Florida and elsewhere (Fig. 10.5). These interactions are mutually beneficial. The K–12 teachers and students benefit from the knowledge and experience of the early-career scientist, and the graduate students report enhanced communication skills and a sense of increased social responsibility – that is, giving back to society via these interactions.

Concluding Comments

Mentoring and role models are fundamentally important to Broader Impacts and should be carefully developed as part of NSF proposals. These activities can be embedded within the research plan, such as more active recruitment of a diverse lab or project group. In so doing, these activities represent value added and do not require additional funds from what typically are already stretched budgets. However, if mentoring and role models are activities that the scientist wants to further develop in a meaningful way, then numerous potential opportunities exist for additional resources from NSF via targeted programs.

The ethical foundation of social responsibility embodies the notion of effective mentoring and role models for people underrepresented in STEM. The culture has shifted at NSF so that broadening participation will remain a priority for the future. Any proposal, particularly those that are larger and more collaborative, will be judged more competitive if diversity, equity, inclusion, effective mentoring, and role models are part of the project.

11 Formal K–12 Education and Partners

> From my many close contacts with outstanding U.S. teachers, I have come to deeply appreciate their wisdom.
>
> Bruce Alberts (2013b: 249)

Code Red Drills

During the 2015–2016 academic year I was on sabbatical as a visiting scientist in the Santa Cruz County (California) Office of Education (SCCOE). I worked with colleagues involved in STEM teaching and learning. This included activities such as professional development, school and classroom visits, talks to civic leaders at Rotary and school board meetings, science fairs, and implementing place-based learning about fossils in elementary schools. During this year I came to better understand the culture of K–12 education. From this first-hand experience came a deep respect and admiration for the dedication these educators have to their profession and students, and the hard work they do.

One thing I had never experienced before was a code red drill, which we did twice during my time at SCCOE. A code red drill indicates an active threat, like a shooter, in the building. During the drill, we all quickly went into interior rooms, barricaded the doors, turned off lights, and hid behind desks. It is a sad commentary on our society that in addition to everything else educators do, they also need to prepare for life-threatening events that potentially occur in the workplace. While schools are not the only targets for shooters, they nevertheless are too frequently chosen as the places to instill terror and tragedy in communities.

Introduction

Formal K–12 audiences are fertile fields for meaningful and innovative Broader Impacts. Any researcher wanting to effectively impact this field must first understand some fundamentals of formal education. Furthermore, the divide between K–12 and higher education, and their respective cultures, is likewise great. Thus, moving outside of our comfort zone into the K–12 world can be a challenge for those accustomed to the ivory tower of higher education.

Many academics are either clueless about K–12 outreach, or simply do not care to get involved. It is too much trouble when they feel that they should be doing more important things related to their jobs. Periodically over the years I have contacted school administrators and suggested programs that are out of their context and not relevant to the needs of their schools. It is therefore not surprising that I have typically not been successful. The few times that I have gone into the local schools to do classroom presentations have been rewarding, but mostly "one-off," with little or no sustained activity. After having worked for a year in a school district (see the chapter's opening anecdote), it is clear that educators have more pressing concerns to contend with, rather than accommodating the whims of ivory-tower academics.

If we are to make a strategic impact on society, for several reasons the audience for this focus should be K–12. First, with more than 50.7 million students, 3 million teachers, and almost 100,000 public schools in the United States (NCES, 2017), just in terms of sheer numbers, the potential impact is enormous. Second, the students currently in K–12 are the future of our country and society; they are the next generation. Nevertheless, in the United States we do not sufficiently value K–12 education. We have lost sight of the importance of this fundamental investment in the future success of our nation. Although the challenges to improving K–12 education are enormous, Broader Impacts can contribute to this segment of society in a meaningful way.

There are many model programs involving scientist–teacher outreach and partnerships in K–12 education. In this chapter we review some of the background and what is needed to effectively develop these partnerships, present some case studies, and general best practices. Educators typically divide the broad field of education into two worlds: formal and informal learning. This chapter focuses on formal education, Chapter 12 focuses on higher education, and Chapter 13 focuses on informal education. Although "science" and "STEM" are sometimes used interchangeably in this book, the concept of STEM has taken hold in education. The latter is mostly used here unless specific reference is being made to science.

Don't Underestimate Serendipity

In 2011 Gary Bloom, then school superintendent (now retired) in Santa Cruz, California and his wife, Katy, took a vacation to Panama where they volunteered working in Carlos Jaramillo's paleobotany laboratory at STRI (Smithsonian Tropical Research Institute). Gary is fascinated with fossils and paleontology. After their positive experience working with Carlos, Gary wanted teachers from his district to have similar, authentic international research experiences. Carlos recommended that Gary contact me, and our subsequent collaboration led to the funding of three

Fig. 11.1 Scientists and teachers looking for fossils along the Panama Canal, 2013 (University of Florida photo).

NSF projects for teacher involvement collecting fossils along and adjacent to the Panama Canal to understand ancient life in the New World tropics (Fig. 11.1), as will be discussed further in this chapter.

This initial, unanticipated collaboration has led to many positive, impactful outcomes, including the ongoing scientist–teacher partnership, role model and classroom visits, and my sabbatical in Santa Cruz. It is likely that other researchers have had similar experiences resulting in successful K–12 outreach and partnerships.

The K–12 Environment: Expectations and Realities

> Unfortunately, teachers and administrators are challenged daily by issues related to student test scores; inadequate facilities; parent concerns; drop-out rates; student mobility; needs of at-risk and disadvantaged students; students who speak English as a second language; and vast socioeconomic, racial, and ethnic diversity.
>
> Moreno (2005: 30)

The culture and work environment within the K–12 profession has many unique aspects, as exemplified by the quote above. Many of these aspects (Inset 11.1) are not familiar to scientists from higher education wanting to develop partnerships (Moreno, 2005; Tanner et al., 2003; Tomanek, 2005). In order to develop effective partnerships, it is helpful for scientists to better understand the K–12 environment. Within this framework, many kinds of activities exist, including science fairs, after-school and summer programs, teacher professional development, and the holy grail – classroom visits and STEM content lessons.

First, we need to fully understand that teachers have at least as many demands on their time as scientists working on research or academics teaching in higher

> ### Some K–12 terms and concepts
>
> 5E lesson plans
> Active learning
> Assessment and evaluation
> Authentic experiences
> Inquiry-based learning
> In-service
> NGSS (Next Generation Science
>   Standards)
> PD (professional development)
> Problem-based learning
> Project-based learning
> Scaffolding

Inset 11.1

education. We all think that we are busy, but teachers are busier and with more structured, daily activities that must be done. A proposal from a scientist to come into the classroom to present a talk or lesson thus may or may not be greeted with enthusiasm by the teacher. The reasons for this reception are varied. Scientists lack specific training in K–12 pedagogy, classroom management, and how to communicate effectively with a younger, formal audience. Teachers oftentimes have their lesson plans and curriculum structured in a certain way and developed far in advance (e. g., months). Therefore, an offer to present STEM content to a class next week or even next month might be out of context with a carefully planned instructional program. Likewise, many teachers' performance appraisals are linked to student achievement, and therefore many have to "teach to the test" – that is, the statewide standardized tests administered near the end of the school year. The impact of these demands on teachers is stressful (Overman, 2018). In Florida, for example, standardized testing resulting from the 2001 No Child Left Behind Act (NCLB, 2001) includes the FCAT (Florida Comprehensive Assessment Test), FCAT 2.0, and related assessments (FCAT, 2018). While these tests may generally be unfamiliar to scientists, teachers are quite focused on these as they relate to curriculum development and their own performance appraisals.

Another major consideration is STEM learning standards, which is a system of performance expectations that guides curriculum development, teaching, learning, and assessment. In the United States, national science standards have existed for a long time (NRC, 1996), but have received an inconsistent welcome depending upon the state. Educational leaders understand the importance of standards for the economy, workforce development, and society. Traditionally, many standards have focused on science content and the rote memorization of facts. Educators also realize, however, that memorizing content is only part of understanding STEM. Thus, how science is done, or the nature of science (NOS) is likewise important, as are connections between different disciplines. The National Academy of Sciences (NAS, 2012) published a landmark book, *A Framework for K–12 Science Education*. This seminal work led to the revisions of many state standards, as well as the creation of the Next Generation Science Standards (NGSS, 2013) that moved STEM learning to a new level of inquiry and integration in the United States. As of 2018, 19 states and the District of Columbia have adopted the NGSS, with other states implementing similar, or opting to keep their own, standards (Fig. 11.2).

Although only being available for less than a decade, the NGSS and similar standards have been widely adopted in the United States (Fig. 11.2). This system represents the future of learning and performance expectations in K–12. Given the

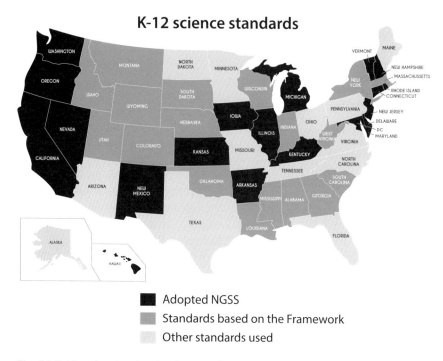

K-12 science standards

Adopted NGSS

Standards based on the Framework

Other standards used

Fig. 11.2 Map showing the distribution of NGSS implementation, by state (NSTA, 2018a).

student populations of the states that have adopted NGSS, these standards affect slightly more than one-third (36 percent) of US students. Also considering the other states that have adopted similar standards, the NGSS are either directly or indirectly impacting more than half (55 percent) of K–12 students in the United States (NSTA, 2018a).

The NGSS have a three-pronged approach that emphasizes three-dimensional learning, including: (1) disciplinary core ideas (content knowledge); (2) cross-cutting concepts; and (3) STEM practices (Krajcik, 2015; Fig. 11.3). Within this framework, the NGSS have learning progressions that build knowledge from one grade to the next (Duncan & Rivet, 2013). A major innovation of NGSS is the special emphasis placed on STEM learning in elementary school (grades K–5). Although teaching science to young learners is fundamentally important (Cook, 2018; Froschauer, 2016), for a variety of reasons this aspiration is a challenge because: (1) in some elementary schools and districts, the previous emphasis was focused on the "three Rs" (reading, "riting," and "rithmetic"), or more recently the Common Core (2018), which meant that teachers were not encouraged, or did not have sufficient time, to teach STEM, even if they wanted to do so; and (2) many K–5 teachers were not formally trained in STEM. These teachers thus lacked the self-efficacy to implement meaningful lesson plans. In summary, standards are a big deal, and their importance in the K–12 environment is even further enhanced by the implementation of NGSS.

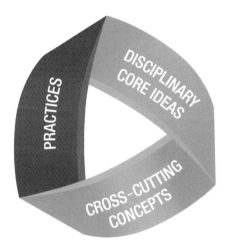

Fig. 11.3 The model of 3D learning that represents the core of NGSS. In one interpretation, Disciplinary Core Ideas are "what students know," Cross-cutting Concepts are "how students think," and Practices are "what students do" (NGSS graphic, with permission).

Therefore, if scientists want to work with teachers, they need to have at least a general familiarity with the relevant standards. It is advisable to present your ideas for lesson content to the teacher having aligned them to existing standards. This should be an instant hook if the teacher is otherwise interested. The point of this discussion is not that it is difficult to work with teachers, but that those wanting to partner within K–12 need to know what teachers need and expect (Moreno, 2005). There are many other reasons why teachers like having scientists in the classroom, including that scientists: (1) can present authentic research; (2) work with real objects and relevant data; (3) can provide role models (particularly from near-peer undergraduate and graduate students); and (4) can discuss career options. All of these are compelling and enhance K–12 teaching and learning. In addition, academics can lead professional development (PD) workshops for the teachers where they can update STEM content knowledge and practices. Many colleges and universities host summer teacher PD institutes (e.g., CPET, 2018; Box 11.1).

Teacher Professional Development

Continuing education and training are important components of teachers' careers and their professional cultures. Although the specifics vary by state and individual school district, all expect some level of PD from their teachers. A certain number of "in-service credits" are required for recertification and also can relate to advancement. Teachers lament the traditional PDs in which they sit in a classroom and listen to an expert drone on about a subject that is of marginal interest or importance to them. These kinds of PDs are disparagingly termed as "sit and get" (Walker, 2013), where the teachers withstand the PD in order to receive the in-service credits.

For a variety of reasons, other kinds of PD that cover topics of current interest and promote more active learning tend to be more popular. Educational administrators

Box 11.1 Fossil Horse Evolution Lesson Plan Development

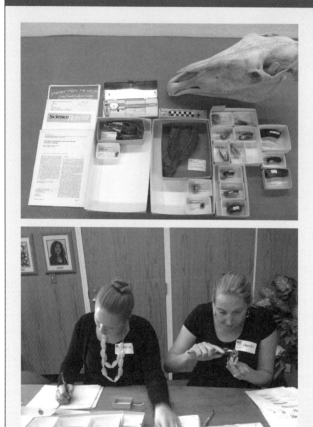

Fig. 1 Top: Study materials used for fossil horse evolution lesson planning (FLMNH, Jeff Gage photo). Bottom: Teachers developing the lesson plan based on the study materials (Julie Bokor photo).

Every summer, CPET (Center for Precollegiate Education and Training) at the University of Florida hosts varied programs of K–12 teacher PD, including workshops and institutes. Within the framework of an evolution- and genetics-themed PD, in 2014 I led a half-day session on fossil horse evolution. After a brief introduction about the importance of fossil horses and a tour of our research collections, the teachers were tasked with developing a lesson plan that they would implement in their classrooms. The deliverables and outcomes of this workshop exceeded expectations. The teachers developed excellent lesson plans related to fossil horses as examples of evolution (PaleoTEACH, 2018). With their background on what was needed for the K–12 classroom, they were

Box 11.1 (cont.)

far more creative than I would have been. In the first lesson, the teachers use fossil teeth spanning the classic 55 million year record of fossil horses in North America. After some scaffolding about the general climatic regimes of the Cenozoic, ranging from the 55 million-year-old Eocene to the 10,000-year-old Pleistocene, students measure fossil horse teeth, plot increase in tooth crown height, and then make interpretations relative to climate changes during this same interval (Bokor et al., 2016). In the second lesson plan, students use photographs to study the variation of dental enamel pattern within a single "paleo-population" (MacFadden, 1989) of the 18 million-year-old Miocene fossil horse *Parahipppus leonensis* from Florida. In the third lesson, after reading a relevant paper on fossil horse displays in museums (MacFadden et al., 2012), students analyze how patterns of evolution would be better displayed in exhibits developed in the future. These lesson plans, many of which follow the 5E model (Krajcik, 2015), have been implemented mostly in high-school biology classes, although some have been adapted for middle schools (e.g., in California, Florida, Michigan, and Oregon).

(e.g., NGSS and STEM coaches) are tasked with ensuring that sufficient PD opportunities exist for teachers to develop the necessary tools to implement NGSS and three-dimensional learning in the classroom (Krajcik, 2015). During a PD for all STEM teachers in the Pajaro Valley Unified School District, Santa Cruz County, California, I was asked to present a talk on a scientist's view of understanding phenomena, which is of current interest in K–12, particularly with the emphasis on NGSS (Krajcik, 2015; Metz, 2018; STEM Teaching Tools, 2018).

Whether it be phenomena or another topic, any novel approaches or fresh perspectives are welcome in PD. As described above, scientists wanting to present teacher PD must understand the culture and motivations of the teachers. If materials are carefully prepared and communicated (with minimal jargon), the audience is typically grateful for a presentation by an enthusiastic scientist. Topics of current interest, such as climate change, evolution, robotics, and 3D-printing, are particularly well received.

Teachers also are expected to participate in PD via attending professional meetings. Given the budget realities, this is often done at their own expense. While many states have teacher- and educator-focused professional meetings, by far the largest meeting that appeals to this community is that of the NSTA (National Science Teaching Association). Thus, for scientists wanting to partner with teachers in meaningful ways, attendance at these annual teacher conferences facilitates networking and building collaborations. Likewise, if the scientist wants to make a presentation, it is best to collaborate with a teacher, one who can frame the presentation for the audience.

Fig. 11.4 UF graduate student Sean Moran talking about 3D-printed fossils at a themed session on the Great American Biotic Interchange (GABI) at the NSTA annual conference in Atlanta, Georgia, March 2018 (author photo).

Case Studies Aligned with NSF K–12 Initiatives

The National Science Foundation has a variety of programs that support the engagement of K–12 teachers within the process of research. Some of NSF's research directorates have specific programs geared toward education that can focus on K–12 teachers or their students as the target audience. A particularly successful program was GEOED, in which active scientists could promote the learning of content in the geosciences. Many practicing geoscience researchers took advantage of these programs. Unfortunately, the GEOED program ended in 2011 (NSF, 2010c). It left a void for scientists wanting to impact education at scales larger than could be accommodated within the Broader Impacts plan embedded within a research proposal.

The NSF's Division of Research on Learning (DRL) within the Education and Human Resources (EHR) Directorate is devoted to advancing understanding and knowledge of teaching and learning of STEM. Several programs have specifically focused on K–12 teacher partnerships. For example, the GK–12 program (AAAS, 2018b), which provided the last cycle of funding in 2010 (NSF, 2009e), placed STEM graduate students as partners with teachers in the K–12 classroom. Having scientists in the K–12 classroom is not a new idea (Rudolph, 2002), and although many challenges exist with this practice there also are potentially profound benefits.

Ongoing programs at NSF, such as DRK–12 (NSF, 2017t), provide funding for innovative K–12 curriculum development and studies of how children learn. The problem for scientists, however, is that most of us do not know how to formally do learning research, which has its own culture and practices. Likewise, evaluation and assessment (Chapter 18) are fundamental components of successful projects in DRK–12 and similar programs; these are also best done by experts in this field. In the case of a STEM scientist wanting to compete within programs such as DRK–12, the best practice is to collaborate with a learning researcher who can frame the theoretical and conceptual frameworks and formulate learning hypotheses in ways that align with these programs.

RET (Research Experiences for Teachers)

Since its beginning, NSF has understood the importance of reaching out to science and engineering teachers by involving them in active research. Silverstein (2008) provides evidence that K–12 teachers' involvement in research positively impacts student achievement in science. Despite this long-standing commitment to teacher engagement in science (and STEM), NSF's RET program is not on the minds of most proposers (NSF, 2012b), and is less well known than the REU program (Research Experiences for Undergraduates; NSF, 2017e). If one substitutes teachers for undergraduates as the target audience, then these programs have similarities – that is, the end goal to foster STEM learning through active participation in authentic research.

We did not initially involve K–12 teachers in our Panama PIRE (2017) project, but after two supplements to expand the scope to this audience we were encouraged by an NSF program officer to submit a RET. We were unfamiliar with this program; indeed, it is not on the radar of most principal investigators (PIs) in the geosciences. The results of our Panama RET, called the GABI (Great American Biotic Interchange), were successful in many respects (Inset 11.2; Fig. 11.5). It is likely that these training and PD projects compare favorably with outcomes from NSF research projects of similar scope and magnitude. In general, few RETs are submitted each year in favor of the more popular REUs. Similar to the model of REUs, RETs can be supplements to existing grants and involve a few teachers each year, typically during the summer. Proposers also can submit requests for RET site grants that can support a cohort of teachers, such as during a summer research institute, typically on a college or university campus working in a research laboratory or doing field work.

GABI RET metrics and products

200 scientist–teacher collaborations

55 teachers and educators spanning K–12 from six states

30 K–12 classroom visits

24 scientists (PhDs, undergraduate, graduate, and interns)

24 professional talks and posters

15 public lectures and talks

12 peer-reviewed papers published

2 project websites developed

Inset 11.2

Certain other commonalities exist between REUs and RETs. Both are expected to address an interesting research topic. Some proposers think of these programs as light on the research, which is supposedly offset by the involvement of either the undergraduates or teachers. Although funding success rates in some REU/RET programs can be relatively high (about 25 percent), it is nevertheless a wise strategy to push innovation in both the research and training programs. Otherwise, many features of standard research proposals also pertain to RETs – for example, collaborations, presentations at professional meetings and conferences, and the production of peer-reviewed research. Another expectation of a RET project is that teachers

Fig. 11.5 GABI RET project-ending summit held in Ghost Ranch, New Mexico, July 2017. Depicted here are the majority of the scientists, teachers, university students, and field interns who participated on this project from 2012 to 2017. (FLMNH, Jeff Gage photo).

should develop lesson plans as deliverables, optimally with wide distribution among the community (e.g., posted on a website or communicated at meetings such as those of NSTA).

It has been our experience working with teachers that, unlike undergraduates who come back to college after the summer, it is challenging to keep teachers engaged in the research project during the school year; they are simply too busy with their day jobs. Thus, projects that want this kind of continued collaboration as an outcome would likely have to schedule, or concentrate, these activities during the summer. This was also a challenge with our GABI RET (2017) because the teachers were from different states and did not live near an active research laboratory affiliated with our project. While there are advantages to the distributed model (Hubenthal & Judge, 2013), there are trade-offs as well. As our summative evaluation of the GABI RET demonstrated (Chapter 10), working with teachers can be a rewarding experience for all people involved, whether they are the scientists, teachers, or students.

ITEST (Integrated Technology Experiences for Students and Teachers)

The long-standing ITEST program (NSF, 2017g) is dedicated to enhancing the STEM workforce. In order for the next generation to succeed in this regard, ITEST promotes learning research on workforce development, acquisition of content knowledge, and twenty-first-century skills such as critical thinking (NAP, 2012). Funded by NSF's education directorate, ITEST programs can include either formal or informal learning environments, or both. Critical components also include the application of novel technologies within the respective learning environment. If these technologies do not exist within the collaborating institutions, then ITEST can help acquire these for the project.

The application of 3D-printing is becoming universal in STEM and other domains in society (e.g., medicine). Applications within the geosciences (Hasiuk, 2014) and paleontology (Tapanila & Rahman, 2016) have demonstrated the potential of this technology within our discipline as well. Fossils and paleontology are a gateway to learning various aspects of STEM and have immense potential in classrooms and maker spaces, for example. An added benefit of making 3D models is that many fossils are fragile, rare, or otherwise unavailable (e.g., still being studied) for distribution on a broad scale to these kinds of learning venues. Seeing this opportunity, we received funding for a three-year ITEST project entitled "iDigFossils" (2017). In this project, more than 50 teachers in middle and high schools in different states participated in the development of learning modules and NGSS-aligned lesson plans that promote the integration of STEM through the lens of fossils and paleontology. During the summers we held iDigFossils institutes in which scientists and teachers came together for professional development. Presentations were made on the background and content of paleontology, including those of current social relevance such as evidence of climate change and evolution in Deep Time. In the early phase of the project, teachers received desktop 3D scanners and printers for deployment in their classrooms. Given that these technologies were mostly new to K–12, our PD was geared toward building the teachers' self-efficacy with these tools and technologies. Outcomes have primarily included the development of 5E lesson plans related to STEM (Krajcik, 2015).

Best Practices and Resources

> Working with K–12 schools is not like crop-dusting—you can't just sprinkle information around and go away.
>
> Mary Margaret Welch, Mercer Island High School,
> Washington (in Dolan & Tanner, 2005)

Given the large audience, partnering with K–12 schools and teachers is potentially of great benefit for society. While it is oftentimes best to start with the teachers "in the trenches," buy-in and support from higher-level educators (i.e., principals and superintendents) are also advisable as a pathway to success. Likewise, individual schools and districts can oftentimes have different priorities; some will have more pressing concerns to deal with. Other schools will welcome partnerships with scientists as a way to enrich the curriculum; this is particularly true if the scientist comes with an understanding of the needs of the teachers, such as standards-based learning. Following up on the quote above, effective teacher PD is best considered as a process between partners, not a "one-off" parachute-drop into the classroom.

Depending upon the school district, a background check, fingerprinting, and possibly volunteer training (e.g., working with minors [students]) may be required. In other school districts, classroom guests will likely need to sign in and out at the main office during each visit. Once you are in the classroom, it then becomes a special experience worth the effort. Few joys are greater in education than engaged kids excited to learn about science, such as through fossils.

In terms of what matters to K–12 educators, demonstrated improvement in student achievement is the gold standard. It is therefore relevant that studies (Silverstein, 2008) show a positive correlation between teacher involvement in scientific research and student achievement. In addition, any activities that contribute to teacher PD, mentoring, or career options for students will likewise typically be considered favorably. It is also important to keep some basic quantitative data on these activities, such as number of teachers served by PD, schools (name) and classroom (and grade levels) visited, or number of students taught. These data will likely become of value in reporting Broader Impacts activities to NSF and when demonstrating prior experience while preparing new proposals.

Although it may not be obvious from the beginning, it is a wise idea to do the IRB (e.g., UF IRB, 2017) approval protocol sooner rather than later. STEM researchers typically do not conduct formal learning/behavioral research that would require approval to study human subjects. Nevertheless, if the participants are surveyed, and any of their responses are reported (e.g., at a professional meeting), then it is recommended that IRB be done.

Scientists new to K–12 outreach do not have to reinvent the wheel. Many resources exist that provide both the conceptual foundations that are important to the teachers, but also practical guides to implementing successful collaborations. A basic selection of these is as follows:

- For general tips and best practices on working with K–12 teachers, see resources such as Moreno (2005).
- For background on national learning standards (performance expectations), see NAS (2012) and NGSS (2013). Note that for states that have not adopted these federal standards, individual standards can be accessed by state (e.g., CPALMS [2018] for Florida).
- For effective professional development for K–12 STEM, see the widely cited paper by Garet et al. (2001) and the Willcuts (2009) report, although many other important references can be found in the professional literature.

Concluding Comments

the single most important thing that practicing scientists in any field can do is reach out to teachers and students at a local high school.

Erin Coffey, Trinity School, New York (2018)

The opportunity to present in a K–12 classroom is a special privilege. Any scientist going into K–12 outreach with the mindset that they are doing the students and teachers a favor by interacting with them ought to find another activity on which to spend their time. In addition to teacher PD and classroom visits, other means of interacting with K–12 audiences also exist, including after-school activities, science fairs (Yoho, 2015), and festivals. Any of these activities has the potential to make a significant impact, not just for compliance with NSF expectations, but for the benefit of society as well.

Although it might be anathema to most aspiring academics, embedding scientists in the classroom is not a new idea (Mervis, 2010; Rudolph, 2002). A paradox exists in modern society in the United States. In many disciplines, particularly in the pure sciences, the supply of new PhDs is far greater than the demand for jobs in higher education and the research sector (Fig. 1.5; Lohr, 2017). On the other hand, school districts – particularly in high-needs areas – cannot effectively recruit and retain qualified science teachers. The National Science Foundation flirted with this paradox with the GK–12 program, but did not go far enough to actually encourage PhDs to opt for careers in K–12. Some PhDs leave research and have successfully pursued rewarding careers in K–12 (Mervis, 2010; Walker, 2018). It is unfortunate that this career pathway is not more firmly embraced by scientists and K–12 educators; it potentially would transform US workforce development and K–16 education in the twenty-first century.

12 Higher Education

Was It the Luck of the Draw?

In 2011 I was invited to participate on a proposal submitted to NSF's Noyce Scholarship Program in the Division of Undergraduate Education. Our collaboration among three colleges on UF's campus (Liberal Arts and Sciences, Education, and Museum) was aimed at undergraduate STEM majors who were part of the UF Teach program (UF Teach, 2018). The goal of this program is to encourage undergraduate STEM majors to pursue careers in K–12 education after graduation. My part involved implementation of a museum internship component of the learning experience. The students would take a summer course with me entitled "Informal STEM Practice" and receive a stipend in return for participating in the internship experience. The reviews of our NSF proposal were mediocre (5 Good, 1 Fair ratings); the panel did not judge it to be competitive, and therefore, not surprisingly, it was not funded. During the next year, the original PI on the proposal took a job at another university and we were left to decide what to do with this project. We made some minor tweaks to the proposal based on the panel review, and resubmitted the proposal largely unchanged from the previous version. To my great surprise, the second proposal received significantly higher reviews (2 Excellent, 1 Very Good, 2 Good ratings) and it was funded.

This example raises the question of whether the composition and dynamic of the review panel, which oftentimes changes from year to year, may have played a part in the higher rankings during the second submission. Or, perhaps in comparison to the set of proposals reviewed in the second panel, our proposal was judged to be better. Although I was skeptical, there also is the other possibility that the minor changes that we made during the second round resulted in a significant improvement to the proposal. We will never know for sure what factors influenced the successful funding of our Noyce Scholarship proposal, but the question remains about the consistency of the panel review process at NSF.

Introduction

Some educators in the United States refer to formal education as K–16, which implies a seamless transition between grades 12 (high-school senior year) and 13 (college freshman year). For a variety of reasons, however, this transition is far less seamless than any other in this supposed K–16 continuum. In particular, this potentially rocky transition relates to the different cultures and expectations of K–12 teachers versus "grades" 13–16 professors, and how the students that they teach learn. It is for this reason that two separate chapters are presented on formal education.

Viewing the entire formal educational continuum as K–12, or K–16, is also admittedly simplistic; there are several complexities to consider further in this model, including: (1) the distinct boundaries between grades 12 and 13 are being softened as increasingly opportunities now exist for high-school students to be dual-enrolled in community colleges, while also taking courses in high schools – at the same time, AP (Advanced Placement) classes likewise soften the K–12/13–16 divide; and (2) many professional jobs and careers now expect additional education past the undergraduate Bachelor's degree. As such, this continuum might be better considered in some professions as K–18 (Masters) or K–21 (PhD).

NSF and Higher Education

> to award . . . scholarships and graduate fellowships in the mathematical, physical, medical, biological, engineering, and other sciences.
>
> National Science Foundation Act of 1950,
> Public Law 81-507 (1950: section 3[4]).

The importance of higher education has been integral to the National Science Foundation's (NSF) mission since its inception in 1950. The first year of funding of proposals in 1952 included 535 graduate fellowships (NSF, 1952). Whereas certain programs and initiatives are ephemeral at NSF, support for graduate fellows has endured for almost 70 years and awarded more than 46,500 Graduate Research Fellowships (GRFs). Former graduate fellows have included many successful leaders in science and technology, including, for example, geophysicist Marcia McNutt, biologist E. O. Wilson, Google co-founder Sergey Brin, and politicans, including former US Secretary of Energy Steven Chu (NSF, 2018d).

Currently NSF directly supports higher education primarily through two divisions within the Education Directorate: DUE (Division of Undergraduate Education) and DGE (Division of Graduate Education). Other directorates and programs also support higher education and workforce development via funded requests embedded within research projects, such as for undergraduate and graduate research assistantships. Taken together, this sustained support for scientific research experiences for students in higher education is an enduring hallmark of NSF.

Don't Forget Community Colleges

> Community colleges play a crucial role in American higher education . . . . It is well documented that community colleges serve a large proportion of minority, first-generation, low-income, and adult students.
>
> Ma and Baum (2016)

While much of NSF's investment in higher education is focused on four-year colleges and universities and graduate schools, two-year "community" colleges that award Associate degrees are also fundamentally important. These colleges, sometimes referred to as junior colleges, began in the United States during the first part of the twentieth century, and expanded widely after World War II. They have typically fulfilled a vocational and technical niche, focused on workforce preparation (Ratcliff, 2018; Trainor, 2015).

In the fall of 2014, 42 percent of all (full-time and part-time) undergraduates in the United States were enrolled across more than 1,000 community colleges (NSF, 2014b). Benefits of the community college are many, including open admissions and lower tuition than many four-year colleges. More so than many four-year colleges, community colleges have a higher percentage of underrepresented (Smith, 2018), low-income, and older students (Ma & Baum, 2016). With regard to underserved minorities, community colleges also have traditionally offered training for many vocational and technical fields typically not supported by four-year institutions. Community colleges also have dual enrollment for high-school students. Although the Associate degree is the norm, some community colleges also award Bachelor's degrees (Santa Fe College, 2018) in fields that either complement, or do not compete with, nearby universities, and oftentimes allow graduates to immediately enter the workforce.

Given the unique characteristics and diverse demographics of community colleges, these institutions are potentially fertile ground for Broader Impacts from a variety of perspectives, including developing partnerships, workforce development, undergraduate engagement in research, and broadening participation. Many NSF programs have emphasis on involvement of community colleges. These include the: (1) Tribal Colleges and Universities Program (NSF, 2018e); (2) REUs (Research Experiences for Undergraduates) such as the nanotechnology site project (NSF, 2017u); (3) Advanced Technological Engineering, or ATE (NSF, 2017v), focused on preparing the technical workforce; and (4) Community College Innovation Challenge, or CCIC (NSF, 2018f). As part of Broader Impacts plans, investigators from universities and other research institutions can also embed partnerships with community colleges in their proposals.

Recent Innovations: MOOCs

> You should consider developing a MOOC.
>
> NSF program officer, *c.*2014

No discussion about innovations in higher education during the twenty-first century would be complete with a discussion of MOOCs, or Massive Open Online Courses. The potential for distance-learning, open education, and engaging large potential audiences embody the concept of MOOCs. The MOOC movement has its roots dating back to a study in 2000 at MIT (Massachusetts Institute of Technology) to determine how curriculum could be delivered over the internet. This ultimately led to MIT instituting OCW, or OpenCourseWare, which has academic content for ~2000 of its courses available online (Carson, 2011; MIT, 2018).

Other initiatives have either embedded existing course content (modified for e-delivery) or developed stand-alone MOOCs. They offer a range of relevant topics for college students, other academics, and also lifelong and leisure learners (my brother-in-law, a retired meteorologist, has taken many MOOCs). Over the past decade a plethora of consortia, or independent academic entities, has emerged in the MOOC market, mostly from the United States, but from other countries as well. As of 2018, about 9,400 MOOCs are being offered by 800 universities and reach 78 million students (Lederman, 2018). In addition to MIT, other big players in this field include Coursera (2018a) and edX (2018). MOOCs may have followed a fad trend ("hype cycle," *sensu* Yang, 2013), and it remains to be seen how this method of e-learning will continue into the future.

Advantages of this mode of e-learning include massive numbers of partici-pants. When Stanford University launched three new courses in 2011, one, "Introduction to AI" (artificial intelligence), enrolled 160,000 students (Pérez-Peña, 2012). While this is an outlier, many MOOCs reach thousands, or tens of thousands, of students. Jordan (2015) observes that a typical MOOC might have 25,000 students enrolled. Another positive impact is access, where large num-bers of participants are enrolled in MOOCs from developing countries. On the down side, MOOCs require lots of time to develop before they are delivered, and managing student input and instructor access is challenging. They are also expensive to develop, with costs ranging from $39,000 to $325,000 (Stainsbury, 2015). Dino 101, hosted by Coursera, had an upfront cost of about $250,000 Canadian (Currie, personal communication, 2018). The primary challenge of MOOCs, however, is that retention and completion rates for courses across the industry are low; they typically range around 10 percent (Franceschin, 2016; Jordan, 2015). Hone and El Said (2016) report in their case study that reasons for low completion rates in MOOCs include factors such as instructors' poor engagement (e.g., the students feeling isolated) and content that was not deemed relevant by the students.

Whether or not MOOCs are here to stay (Pope, 2014), they are an interesting twist to cyberenabled learning. Likewise, MOOCs are an innovative way to break down geographic and access barriers and thus reach large numbers of learners. It nevertheless remains to be seen how accessible and effective this platform will be in the future of e-learning, or whether it is just a passing fad (Yang, 2013; Adam, 2019).

Box 12.1 Anatomy of a Dinosaur MOOC

 **DINO 101**

(University of Alberta, Philip Currie permission)

This is a great course! Loved the 3D fossils.

Liam, eight years old, in Coursera (2018b)

It might come as a surprise that an eight-year-old took Dino 101, but another advantage of MOOCs is that they can be accessed by learners of all ages. As Liam likely could tell us, kids like dinosaurs, so a MOOC on this topic seems to be a natural draw. Dr. Philip Currie, a world-renowned expert on dinosaur paleontology, teaches a dinosaur paleontology MOOC via Coursera. This is a 12-week course requiring about 1–2 hours of study per week. It covers a variety of themes related to the dinosaur paleobiology, evolution, and related topics in biology and geology. It also has an online community portal where students can "chat" with other students about the topics of the course. Students can either sign up via Coursera for a certificate, or enroll in the course through the University of Alberta, the latter of which has exams. This university course is taught each semester and typically has about 800 students enrolled; at the same time, a greater number are connecting via Coursera. As of 2018, Dino 101 has been taught for five years and Dr. Currie reflects that:

I am still surprised about how strong it is going, and about the letters I get from all around the world. I initially did not believe that this could replace a university course, but now I am convinced that it is a good way to go for at least basic courses.

Case Examples of Broader Impacts in Higher Education

When Broader Impacts was first implemented in 1997, the default and somewhat unimaginative plan involved the investigator proposing to integrate the content of their research into formal courses. This might have included, for example, a few lectures woven into the fabric of the existing course content. Another solution was to propose that undergraduates would be recruited into their research laboratory. Expectations were otherwise uncertain during the early days of the Broader Impacts merit review criteria; these plans to teach, or directly involve, undergraduates seemed reasonable and fit within the comfort zone of most academicians. Over the past 20 years, however, the landscape and expectations of Broader Impacts have increased with regard to involving participants from higher education. A quarter-century later, researchers proposing to work with students in colleges and universities have to develop a more innovative plan, as well as involve diverse participants. Most of us were guilty of proposing some of these default activities in the early days of Broader Impacts. Nevertheless, Broader Impacts activities related to higher education have likewise evolved and involved both undergraduates and graduate students. Two recent projects are presented below.

Undergraduates: Informal STEM Practice

Related to the anecdote at the beginning of this chapter, in 2012 I began teaching an undergraduate course entitled "Informal STEM Practice" to students in the UF Teach (2018) program, which includes undergraduate STEM majors who opt to teach in K–12 after graduation. Funded by NSF's Noyce Scholarship Program (NSF, 2017w), the students received a modest summer stipend. In return, they (1) developed an internship in an informal STEM institution; and (2) took this undergraduate course for 12 weeks. The internship required 240 hours of work at a local museum or other informal STEM institution. Because many of the students went home during the summer, we delivered the weekly three-hour class both in the physical classroom and via videoconferencing (also see Fig. 15.1). The classes typically had a three-part format with: (1) a talk by an expert on the informal STEM topic of the day; (2) student-led discussions of assigned readings; and (3) a weekly sharing of the students' internship experiences. As part of their homework, students were expected to either blog or write Facebook posts, and also comment on other students' posts. This course, paired with the summer internship experience, was held for five summers. Van Duzor and Sabella (2012) promote the benefits and success of these kinds of internships for pre-service STEM teacher professional development and retention. They highlight the efficacy of these internships at informal STEM institutions in terms of mentoring, best practices, and the students' development of a "sense of place" in the local community. Our students had previously been unaware of the ways in which these kinds of institutions could support and enhance K–12 classroom teaching and learning.

As we will also learn in the next chapter, 95 percent of all learning occurs outside of the formal classroom (Falk & Dierking, 2010). Our UF Teach students had not previously understood the potential impact of informal education on overall learning. Although many of the students in this course ultimately went on to teach STEM, some did not; some looked for jobs in informal STEM institutions, including museums and planetaria. Nevertheless, our course confirmed other studies, such as CalTeach (Whang-Sayson et al., 2017), documenting the value of the Noyce Scholarship Program. The UF Teach students who participated in this program developed a deep appreciation for the K–12 teaching profession.

Graduate Students: Broader Impacts

> Taking the broader impacts class shaped the way I saw my career. I realized that I needed and wanted to do both, publish scientific papers, and reach out to the general public. Making this decision changed the classes I would take during grad school (ended up minoring in science education and technology) and the products I would produce. My CV started having sections with lists of outreach products, invited talks in museums and schools, and papers in both the scientific literature and in education and outreach. That set me apart from other scientists and gave me my own trade mark.
>
> Catalina Pimiento, PhD (zoology), University of Florida (2018 email to author)

Realizing the need among the next generation of academic professionals, and the general lack of awareness, in 2006 I started teaching a graduate seminar on Broader Impacts; this course has now been taught seven times. During the first half of the semester, the classes typically include talks presented by the instructor or guest experts (Inset 12.1; compiled from author's course syllabi). The second half of the semester is devoted to topics chosen and presented by the students.

The original intent of this course was to make it broad enough to appeal to graduate students in any STEM discipline. Despite this intended reach, however, students have been drawn from the natural sciences; those from more distant fields, such as engineering, typically do not take this course despite attempts to market it widely throughout the university. A systematic analysis explaining why this course does not have a broader appeal has not yet been done. Anecdotally, it may be that the students are already occupied with other required courses, or perhaps their major professors do not recommend that they take such a course (MacFadden, 2009).

Our Broader Impacts graduate course requires a final class project as part of their assessment. Each student typically does their own project, or sometimes they work in small groups. However, after much discussion, the 2008 Broader Impacts class decided that they wanted to do a single project together. Their project involved a front-end evaluation (Chapter 18) of museum

Broader Impacts class topics

Core topics
Background, NSF and Broader Impacts
Diversity and broadening participation
Evaluation and assessment
Science communication
Museum exhibits and programs
K–12 Education and outreach
Websites 101
Community (citizen) science

Activity-based classes
Elevator speeches and sparks
Mock NSF panel review
WeDigBio transcription hackathon

Students' topics
3D-printing and K–12 outreach
Agricultural extension networks
Misconceptions and controversies
Social media and blogs
Open-access publishing
Pop-ups at fairs and festivals
STEAM

Inset 12.1

visitors' preferences for different themes that the museum was considering for upcoming temporary exhibits. The students formed a mock evaluation firm and developed a contract with their "client." They then developed a survey that was, after IRB approval, administered to more than 100 visitors to our museum. The team spent more than 100 hours "on the floor" interviewing visitors. They compiled the data and then made a presentation of their findings to the client. Student reviews of this class project placed high value on this shared experience. One of the exhibits that the visitors recommended, *First Colony* – about the settlement of Florida in Spanish colonial times – was ultimately implemented. It then morphed into a permanent exhibition at Government House in St. Augustine, Florida (FLMNH, 2018).

Outcomes of this Broader Impacts course have included short-term and long-term benefits to the students. Along with Dr. Pimiento's quote above, evidence of the efficacy of this class has ranged from successful Broader Impacts plans based on NSF reviews, to feedback about how an innovative outreach plan was the margin of difference in a highly competitive postdoctoral fellowship. Some students in the Broader Impacts class have also published papers on their projects in peer-reviewed journals (e.g., DeSantis, 2009a; 2009b). As such, they have diversified their resumés in preparation for a tight job market. It is also becoming more common at the University of Florida (UF) for graduate students to include a separate chapter in their thesis or dissertation describing their Broader Impacts-like project. Other successful outcomes have included students later successfully implementing their plans and disseminating the results of their projects to different audiences.

Crisis in Graduate Education

For decades the job market in basic science, and more recently STEM, has been tight. It has not gotten any better in recent years; in fact, it seems to have accelerated down a slippery slope. The gold standard – the academic, tenure-track job – is becoming increasingly more difficult to find, particularly in fields that are neither rapidly growing, nor in high need (such as biological science; Fig. 1.5; Lohr, 2017). As such, the supply of PhDs in some disciplines is an order of magnitude greater than the demand for academic positions. Cassuto's (2016) book, *The Graduate School Mess*, discusses this problem within the humanities, but many of the same challenges and problems also extend to the sciences.

Broader Impacts training can potentially help in the tight job market as long as the student is flexible with regard to career options – that is, if they would consider training for careers outside of the classic tenure-track job in academia. Certain "soft skills" acquired during a graduate education are transferable to many different careers, although these are not typically taught in graduate school (*Nature*, 2017). Some of these, such as increased proficiency with science communication, can be gained through Broader Impacts activities. In the future, it would also be beneficial to offer additional training on alternate careers in STEM during graduate students' education. Few universities do this now, but this kind of training seems to be increasing, and will likely move in this direction in the future.

Concluding Comments

Despite being maligned earlier as the unimaginative default activity within academics' comfort zone, plans to involve higher education within scientists' research activities are potentially effective opportunities within a Broader Impacts framework. Likewise, the landscape is changing at NSF and any activity that will broaden participation as part of a Broader Impacts plan will be more competitive. Indeed, it seems that in some programs projects will be significantly more likely to be funded if diversity and related initiatives are woven into the fabric of the project.

The notion that a seamless transition exists between K–12 and 13–16 is a well-known theoretical construct, but in reality the divide is deep, mostly because of fundamentally different cultures within these two professional realms. To some extent, programs such as dual-enrollment and AP courses in high school continue to soften the transition. In addition, the role of community colleges within our society, particularly in the United States, are an important component of higher education.

For scientists who find Broader Impacts activities within higher education rewarding, other opportunities exist to develop these interests in addition to individual research proposals. For example, REU programs (NSF, 2017e) and graduate programs such as the NSF Research Traineeship (NSF, 2018g) have great potential to develop the next generation of professionals within the STEM workforce. It also is surprising that more universities have not implemented Broader Impacts training for the next generation. This seems like a missed opportunity not just for graduate students writing NSF proposals, but also as a soft skill that will position them better in the workforce during their professional careers.

Technology and e-learning will continue to be platforms for innovation and experimentation within higher education. Courses in which students participate in real time in a physical or virtual classroom and then use social media to communicate in learning networks outside of class will continue to broaden access to learning. The advent of MOOCs may signal new opportunities for innovation within some sectors of STEM. All of these programs and activities indicate that higher education is fertile ground for Broader Impacts activities in the twenty-first century.

13 Informal STEM Learning in Museums and Beyond

Does NSF Build Exhibits?

In the late 1980s I was working with a colleague in our museum's education division to build a small exhibit on fossil horse evolution. She suggested that I talk with program officers in NSF's Informal Science Education (ISE) program about potential opportunities to fund our project. I contacted Barbara Butler, then an ISE program officer, and arranged to meet with her during one of my trips to Washington, DC (where NSF was then located).

I naively started the meeting by saying that we were interested in building an exhibit on fossil horses and that I had heard that NSF supports these kinds of projects. After my preamble, Barbara had me in her cross-hairs. From what I can remember, she asked me what the learning goals were for the project, what did we know of the audience demographics, had we done a needs assessment, and what were our plans for evaluation? To say that I was ignorant of these questions is an understatement. I remember Barbara telling me that NSF does not build exhibits, *per se*, but they fund learning research in informal settings like museums. If the research needed an exhibit to answer an interesting learning research question, then building it could be part of the project. She ended the meeting by recommending that I go back and talk with my colleague about these questions and get back with her if we wanted to proceed with a proposal. With these seeming impediments, we never did proceed, and thus NSF did not fund our project.

Despite this rocky start in the field of informal science education, and what likely was a bad first impression on Barbara, things improved thereafter. A decade later she funded our *Fossil Horses in Cyberspace* project as an ISE supplement. Ironically, three decades later, when I was a program officer in ISE, and after Barbara retired from NSF, she served as a proposal reviewer for panels that I led at NSF. If she remembered the first meeting in which we talked about the fossil horse exhibit, she was gracious enough not to bring it up in conversation.

Introduction: The 95 Percent Solution

> School is not where most Americans learn most of their science.
>
> Falk and Dierking (2010)

In a 2010 report titled *Surrounded by Science,* Fenichel and Schweingruber (2010) present the essence of informal science education, or in a more modern context, informal STEM learning (ISL) (NSF, 2017x). Falk and Dierking (2010), two innovative leaders in this domain, emphasize the fact that during a person's average lifespan, about 95 percent of their waking hours are spent outside of the formal classroom, or conversely, 5 percent in formal classroom instruction, and of the latter, only a small fraction is dedicated to STEM (Fig. 13.1). If we want to impact STEM learning, then informal settings, venues, programs, and activities potentially provide the greatest opportunity. People learn all the time, and thus give rise to the concept of "lifelong learners," or in the context here, lifetime STEM.

Informal science education traditionally has been overlooked relative to formal education. There are many reasons why this is the case. Society views formal education, particularly K–12, as a fundamental entitlement, and thus much of the emphasis (but typically not enough resources) has been placed on this kind of learning. Most education colleges in the United States focus on teacher training, counseling, assessment, and educational administration. With few exceptions, such as the Free-Choice Learning program at Oregon State University (OSU, 2018), the academic pursuit of informal science education or STEM learning research is typically relegated to a few professors, or done by professors assigned primarily to formal education. Another related issue is the matter of assessing outcomes. Formal educators are heavily focused on student achievement as important outcomes of learning activities. Thus, the assessment of learning is fundamental to the accountability of formal education. In contrast, learning gains are rarely assessed during an informal STEM activity like a trip to a museum. Because these activities did not account for traditional learning, they were considered less meaningful. Informal educators have understood this lack of accountability and developed a system of six strands

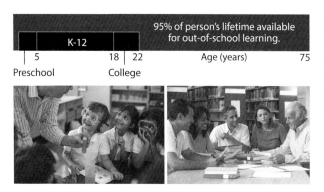

Fig. 13.1 Model showing that most of a person's lifetime, and therefore opportunity to learn, is spent outside of the classroom (Adobe stock photos).

Six strands of informal science

1. **Developing science interest**
   Excitement
   Interest
   Comfort
2. **Understanding scientific knowledge**
3. **Engaging in scientific reasoning**
   Interactivity
   Doing and seeing
   Meaning-making and explanation
   Questioning and predicting
4. **Reflecting on science**
   Self-reflections on learning
5. **Engaging in scientific practices**
   Parent–child interactions
   Specialized science talk
   Scientific tools
   Social group influences
6. **Identifying with science enterprise**
   (Cognitive) agenda
   Prior knowledge and experience
   Personal commitment to action
   Building science identity across
      age and background

Inset 13.1

(Bell et al., 2009; Chapter 9; Inset 13.1) primarily focused on designed and built settings, in which learning and related affective behaviors can be evaluated (Bell et al., 2009).

Fifty million children go to school in the United States (NCES, 2017); an equal number of people visit natural history museums each year (MacFadden et al., 2007). When one also considers other options and opportunities for learning in informal settings, the potential impact on ISL is all-encompassing and cuts across many demographics and society at large. To give a few other examples, although attendance at museums and other "built" venues is an intentional activity that primarily serves a geographically limited audience, the potential reach of ISL from other means of dissemination is extraordinary. Several large airports in the United States have STEM-themed exhibits; for example, dinosaurs are on display at Dallas Fort Worth and Chicago O'Hare (Fig. 13.2). These two airports are among the busiest in North America. In 2017 together they accounted for 147 million passengers (Zhang, 2018), almost three times as many people as visit US natural history museums. These typically are passive "exhibits" that have not been evaluated for impact, but O'Hare had "Dino-Day" with more active themed promotion (Baskas, 2016). While it cannot be disputed that the intent of many of these themed exhibits in "non-traditional" venues is to promote local natural history museums (also see "Curate My Community" in Chapter 3), most of the travelers rushing to catch their flights likely do not learn much, if anything, about science from these exhibits. Turning to another public activity – sports – about 50 million people each year attend college football games (NCAA, 2018). Coincidentally, this number is about the same as those who visit natural history museums. One particularly innovative program has been a short (one-minute) science factoid about the physics of football presented during halftimes at the University of Nebraska by professor Tim Gay (2005; UNL Physics, 2018). Many other opportunities exist to communicate STEM in informal settings, as will be described below.

> Would you rather publish a paper that 10 colleagues cite, and 100 scientists read, or build an exhibit that hundreds of thousands of the public get to enjoy at a museum?

Fig. 13.2 A dinosaur (*Brachiosaurus*) skeleton replica from the Field Museum on display at O'Hare Airport, Chicago (James St. John photo; also see Keoun, 2000).

I ask this question when I start professional development (PD) for faculty, and also to students in my Broader Impacts class. Sometimes faculty want to debate that these are like comparing apples to oranges. In contrast, the students, perhaps with more open minds and less prone to instant debate, tend to better understand this comparison. Informal science, or STEM, is an excellent gateway to potentially impact broad segments of society via communication to the public of research discoveries. Likewise this field potentially is an effective way to benefit society, whether it be through options such as: (1) designed or built environments; (2) non-traditional venues; (3) non-formal and after-school programs; (4) mass media; and (5) fairs and festivals. Once STEM professionals have stepped into this arena, it has been my experience that many find it personally rewarding and tend to remain engaged in informal learning as they cycle through series of research projects.

Box 13.1 Selected Informal STEM Resources and Programs

AISL (Advancing Informal STEM Learning)
is the NSF (2017x) program that funds many kinds of informal science (STEM) education (learning) projects.

CAISE (Center for Advancement of Informal Science Education)
is an independent professional network funded by NSF that advances the field of informal STEM education "by providing infrastructure, resources and connectivity for educators, researchers, evaluators, and other interested stakeholders" (CAISE, 2018).

Connected Science Learning
(2018) is a recently established peer-reviewed, online journal published by the National Science Teaching Association (NSTA) that promotes connections between formal and informal learning.

Institute of Museum and Library Services
(IMLS, 2018) is a federal agency that provides funding for programs, including informal education, within museums and libraries.

Surrounded by Science.
Many books exist that provide support for informal science education. Fenichel and Schweingruber's (2010) book is based on contributions from leaders in this field, mostly in the United States. In contrast to the more scholarly *Learning Science in Informal Environments* (Bell et al., 2009), which is a dense academic book of edited articles, *Surrounded by Science* is a more inviting and accessible book. The latter was used as a text reference in my Informal STEM Practice course (Chapter 12). As with all publications of the National Academies Press, these books can be downloaded as PDFs for free.

Visitor Studies Association
(2017) is a small professional organization focused on improving the understanding of visitor experiences at museums, science centers, and other built environments. It also publishes a professional journal of peer-reviewed articles, *Visitor Studies: Theory, Research, and Practice.*

As Falk and Dierking (2010) emphasized in their 95 percent solution, informal learning can take many forms, be done in many kinds of venues, and occur throughout a person's lifetime. In addition, ISL activities that are part of a comprehensive

Broader Impacts plan have the potential to broaden participation and also enhance public participation in science. This is the first of a series of four chapters related to ISL. Other related activities such as community (citizen) science, the web, social media, online learning, and related topics are presented in Chapters 14–16.

Designed and Built ISL Settings: Beyond Dusty Old Exhibits

When scientists think about informal science education, a good place to start is with designed or built settings. My first free-choice learning experiences as a child were when I went to the American Museum of Natural History (AMNH; Fig. 13.3) to look at the dinosaur exhibits (see Inset 13.2 for nuances of "exhibit" versus "exhibition," Grammarist, 2018). Although I did not first visit it until graduate school, the Natural History Museum in London is another great museum, and like the AMNH, is a venerable and iconic nineteenth-century icon. I've been asked more than once what my favorite exhibit is. "Sense of Wonder" at the Milwaukee Public Museum (Fig. 13.3) is an object-rich exhibition with more than 1,000 specimens on display that embody the essence of what makes a natural history museum interesting. It is small, yet impactful. Other top contenders include the main gallery at the Oxford (UK) Museum of Natural History (Fig. 13.3) and Cuvier's "cabinet of curiosities" in Paris (Fig.13.3). What do these examples have in common? All are specimen- or artifact-rich, and the natural history museum displays objects in a Victorian-like style. It is perhaps not surprising that real specimens are a hook (Benton, 2010); this is one of the prime reasons why people go to natural history museums (Falk & Dierking, 2010). Natural history museums, however, now realize that they cannot perpetuate the disparaging public notion that they are a place to find (mostly) boring dead things on display, or the notion that natural history is an antiquated science relegated to the nineteenth or twentieth centuries (Chicone & Kisel, 2014; Dorfman, 2017; Rader & Cain, 2014).

Exhibit versus exhibition
Although these words are often considered synonyms, for the purist:
*The difference between **exhibit** and **exhibition** is a matter of scale.*
"An exhibit is a public showing of an object – usually work of art or an object meant to educate– or a small collection of objects.
An exhibition is a public showing of a large selection of such objects.
For example, a fossilized dinosaur skeleton in a lobby of a museum is an exhibit, and a collection of dinosaur skeletons in a wing of the museum might be called an exhibition. "

Inset 13.2

A major challenge with natural history museums is that exhibits are very expensive to build. Over the past two decades we have built a half-dozen permanent exhibitions at the FLMNH that have ranged in cost from $500 to almost $2,000 per square foot, the latter for technology-laden displays. Whitemyer (2018; also see Walhimer, 2011) reports that natural history

Fig. 13.3 Object-rich natural history museum exhibitions. Top-left: Exhibition hall at Naturalis Biodiversity Center Leiden (with permission). Bottom-left: Sense of Wonder, Milwaukee Public Museum (Jim Trottier photo). Top-right: Main exhibition gallery, Oxford Museum (Oxford Museum photo). Bottom-right: Gallery of Paleontology and Anatomy (Cuvier's cabinet of curiosities) Paris (Sarah Boessenecker photo).

museum exhibits can range from $75 to more than $800 per square foot. Because of their expense, permanent exhibition halls are typically not renovated for decades (or in some cases, centuries), and thus do not keep up with modern discoveries that advance scientific understanding. Antiquated exhibits also do not keep up with modern learning strategies in which interactivity is a key to visitor engagement. Fossil horse (Family Equidae) exhibits are a good example of potential misconceptions about the pattern of evolution based on outmoded or oversimplified exhibits (Dyehouse, 2011; MacFadden et al., 2012). Older exhibits built at the beginning of the twentieth century depicted fossil horses as a linear array of ancestral and descendant species (Fig. 13.4), supporting the notion of "orthogenesis" or straight-line evolution. Paleontologists, however, have known for a century that horse evolution is more accurately depicted as a complex, or bushy, tree without the fossils being organized in a tidy linear pattern. Thus, many updated fossil horse exhibits now display the evolutionary complexity of this group in their renovated exhibits (Fig. 13.4).

In terms of potential public impact, natural history museums are only the tip of the iceberg. Thus, many other kinds of institutions are popular venues for informal STEM learning: for example, science centers, which were made popular

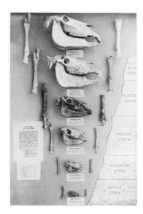

Fig. 13.4 Left: Early twentieth-century depiction of horse evolution as a straight-line (orthogenetic) sequence through time (Lucas, 1927). Right: "A Textbook Case Revisted," the fossil horse exhibit at the AMNH in the twenty-first century. In contrast to the straight-line sequence from the twentieth century, the fossil horse skeletons today are not organized into a tidy unidirectional display, but rather have staggered sizes and are facing opposite directions, deliberately to dispel the idea of orthogenesis or orthogenetic evolution (Benjamin Miller photo).

by venues such as the Exploratorium in San Francisco. The Exploratorium was envisioned by the twentieth-century physicist-turned-teacher Robert Oppenheimer (Allen 2004; Exploratorium, 2018; Rader & Cain 2014). Although there can be some overlap in exhibit and public engagement strategies, relative to natural history museums, science centers have a different focus, most notably that they have traditionally been more experiential and do not typically maintain extensive research collections (Dorfman, 2017). Science center exhibits tend to be more hands-on, tech-laden, and span a broader range of STEM. They also tend to present other STEM topics of societal relevance, such as the development of computer technology, not always covered in conventional natural history museums. Research has also shown that science centers and natural history museums attract different kinds of visitors with different expectations of their anticipated experiences. For example, from the sample analyzed, science centers typically attract larger groups with more children. Likewise, visitors to science centers expected relatively more technology in the exhibits (Korn, 1995). Despite this classic study being done a quarter-century ago, other studies also underscore the importance of understanding specific visitor demographics for a particular institution (Dorfman, 2017; Korn, 2018).

Having noted these differences in institutional missions, however, many science centers – such as those in Baltimore, Maryland, and Fort Worth, Texas – have dinosaurs prominently on display, likely capitalizing on their immense public appeal, particularly to children. Some recent institutions, like the Perot Museum of Nature and Science (2018) in

Fig. 13.5 Visitors to the Butterfly Rainforest at the FLMNH (FLMNH photo).

Dallas, Texas, have blended the specimen-based content of the natural history museum with the interactivity of the more experiential science center. Other designed spaces that promote STEM run the gamut of aquaria, planetaria and observatories, zoos and botanical gardens, outdoor spaces in local nature and environmental centers, and US National Parks (Bell et al., 2009; Fenichel & Schweingruber, 2010). Similar to the differences seen in visitor demographics in natural history museums and science centers, in 2004 our museum opened the Butterfly Rainforest as a public engagement and learning space within our museum. Ranked one of the "Top Things to Do in Gainesville" (Tripadvisor, 2018), we have noticed that visitors come to this exhibit for the immersion experience (Fig. 13.5). A bonus to the bottom-line is that, since it has been opened, the Butterfly Rainforest has helped to boost our annual attendance (FLMNH, 2018; Fig. 8.3).

Open Exhibits and Live Laboratories

Particularly in natural history museums, much of the building space (in some cases more than half) is devoted to "back-of-the-house" activities, such as preparing specimens, collections storage, and research laboratories. Most visitors to these institutions, therefore, do not understand the resources devoted to these activities, mostly because they cannot see them. These spaces and their activities are typically inaccessible to the general public. Nevertheless, the public is fascinated with many of the activities that occur in these spaces. Decades ago we sponsored a "behind-the-scenes" open-house one day per year that would attract about 1,000 visitors from the local community. This interest underscores the general notion that the public is fascinated with both the practice and process of science, the latter sometimes termed "NOS," or nature of science (e.g., McComas & Nouri, 2016; NGSS, 2013; NSTA, 2018b). Capitalizing on this fact, many museums have deliberately built: (1) open storage, in which the public can see actual research collections (IME, 2009; Fig. 13.6); and (2) preparation labs, where the public can see museum staff working. Research collections on display are not unique to natural history museums. One particularly inspiring example is at the New York Historical Society (Fig. 13.6; NYHS, 2017), across the street from the AMNH in New York.

Fig. 13.6 Top: Lapworth Museum of Geology, University of Birmingham (with permission). Middle: Open collections storage on display at the Luce Center, New York Historical Society (Mark B. Schlemmer photo).Bottom-left: FLMNH Centennial Live lab. (FLMNH, Kristen Grace photo).Bottom-right: Fossil lab, which is open to conversations between visitors and volunteer preparators (behind the window), at the Calvert Marine Museum, Solomons, Maryland (author photo).

Take-Home Message: Broader Impacts, Access, and Feasibility

It is likely that hundreds of millions of people visit designed informal learning spaces throughout the world each year. As such, these spaces have the potential to benefit society because of their impact on committed audiences. However, exhibits within these spaces can be very expensive to build, particularly if they are intended to be permanent exhibitions. Scientists wanting to participate in exhibit projects may find shorter-term temporary displays more cost-effective.

With the exception of organized school groups, these venues only serve the segment of society that chooses to spend their time at a museum, science center, or other informal venue. In addition, with the exception of national and state parks located in remote areas, other physical venues are not available, because of lack of geographic proximity, for those in rural, underserved communities. Scientists seeking to broaden participation should thus keep these factors in mind when planning their activities.

Mobile and "Pop-Up" Museums, Fairs, and Festivals

Several challenges exist with the intentional, design-based learning spaces described above, including free-choice, access, and proximity. Falk and Dierking (2010) have demonstrated the value of free-choice learning in these spaces; however, given the limited free time available, the public needs to be motivated to visit museums, science centers, and the like. Even when cultural institutions are nearby, people from certain socioeconomic and ethnic backgrounds disproportionally tend not to go to them (Farrell & Medvedeva, 2010). This is particularly true with underserved minorities living in urban and inner-city areas, as well as rural communities in the United States.

One way to address this disparity is via mobile museums, which, rather than requiring the public to visit the museum, allow the museum to go into the community. The concept of a mobile museum is not a new one (Fig. 13.7); it has been around for more than a century (Rees, 2016). The success of this kind of outreach depends upon the target audience. Although mobile museums potentially reach large audiences that are otherwise underserved by the home institution, with the exception of captive audiences of K–12 schools, the visitor still has to choose freely to visit them.

Just like the concept of non-traditional venues for the built environment, so too is there an opportunity to reach large audiences at public events not typically themed for STEM. We discussed the recent innovation of "pop-up" museums in Chapter 9, but they are also relevant here. Originally started in the art community, pop-ups have spread to other domains, including STEM. Simply put, pop-ups are small exhibits that are, for example, displayed outside under a canopy, that can communicate STEM content. In addition to communicating specific content, Nina Simon, a leader in informal STEM learning innovation (Simon, 2011), believes that pop-ups can serve the community by two-way participatory dialog – that is, they can be "a way to catalyze conversations among diverse people, mediated by their objects." At our museum we have experimented with a variety of pop-ups at local fairs and festivals. These are typically aligned with the science theme of the festival, such as an ichthyology pop-up at a fishing tournament (Fig. 9.3). Starting in 2015, we have been

Fig. 13.7 Left: Bus converted to a mobile museum of the Florida State Museum (now FLMNH), circa mid-1950s. The bus traveled to distant venues, mostly schools, in Florida and reached relatively large K–12 focused audiences, exceeding several tens of thousands (Grobman, 1955; FLMNH Archives photo). Right: Mobile Earth and Environmental "GeoBago" vehicle from the Appalachian State University that conducts outreach in rural North Carolina (Marta Toran photo).

Fig. 13.8 Pop-up about fossil elephants at the peanut festival in Williston, Florida. (FLMNH, Jeff Gage photo).

excavating extinct elephant-like fossils (gomphotheres) from Montbrook, a fossil site near Williston, northern Florida (Hulbert, 2016), located less than an hour's drive from our museum. Associated with a media blitz, at a recent "Peanut Festival" (a significant cash crop in this region), we had a pop-up about Montbrook fossils, focusing on the gomphotheres (Fig. 13.8).

From the point of view of Broader Impacts, pop-ups are a cost-effective and rewarding way to communicate STEM to the public. Relative to permanent institutional venues, they can be relatively easy to design and implement, and their costs are nominal. In contrast to many other projects, such as self-guided museum exhibits or websites, where the users are typically not known, pop-ups engage people directly. Many scientists find that direct communication with their intended target audience can be immensely rewarding. We intend to continue to use pop-ups in remote locations to encourage further participation and engagement.

Box 13.2 USA Science & Engineering Festival

Each year this festival is held during a spring weekend in Washington, DC. In recent years it has included 3,000 hands-on exhibits and more than 350,000 attendees in one weekend (USA Science & Engineering Festival, 2018). It therefore is an excellent way to promote STEM, with large numbers of actively engaged members of the public. Similar to other informal settings, most of the attendees (with some exceptions such as organized K–12 school groups) have freely chosen to visit the festival, so the audiences are typically interested in the exhibitions. This is likely one of the largest weekend STEM festivals, and therefore has the potential for immense reach during a relatively short time interval. On two occasions we sent teams to these festivals to promote our NSF-funded Panama PIRE (2017) and iDigBio (Page et al., 2015) projects.

Fig. 1 Top: Overview of the USA Science & Engineering Festival (USA Science & Engineering Festival, 2018). Our 2012 Panama exhibit display booth below included hands-on active learning and engaged hundreds, if not thousands, of festival-goers. Bottom-left: Two young learners talking in real-time to paleontologists collecting fossils along the Panama Canal. Bottom-right: Graduate student Sean Moran (right) and a festival visitor comparing a modern horse skull to a 20 million-year-old fossil horse collected from the Panama Canal.

Box 13.2 (cont.)

Most science fairs in the United States traditionally focus on K–12 student projects, but venues such as the USA Science & Engineering Festival in Washington, DC, facilitate professionals participating in these kinds of events. Whether they are enormous annual expos or local community events, STEM-themed festivals are an excellent way to directly reach informal learners (Borysiewicz & Buckley, 2014; Durant, 2013). Other kinds of public events like state fairs also attract large audiences in a relatively short period of time. These festivals and fairs therefore provide excellent opportunities for societal impact. In addition to the benefit accrued to the visitors, festivals and fairs also provide the opportunity for scientists to have active engagement with thousands of "the general public" during the festival. In so doing, scientists have the opportunity to hone their science communication skills and present their message for societal benefit.

Some cities sponsor annual science festivals. These events have the potential to attract large audiences during the festival and therefore provide meaningful science outreach within the sponsoring communities. This is exemplified by the annual Los Alamos Science Fest (2018) in New Mexico, a city that was essentially created to develop atomic energy during and after World War II as part of the Manhattan Project. Likewise, university cities, such as Cambridge in Massachusetts (Durant, 2013) or Cambridge in the United Kingdom, sponsor an annual science festival hosted by the University of Cambridge (Fig. 13.9; Cambridge Science Festival, 2018). Started in 1994, this festival has grown from modest beginnings to now involving hundreds of scientists and attracting more than 30,000 attendees. In addition to the Cambridge festival, Borysiewicz and Buckley (2014) report that several hundred festivals worldwide engage millions of people. They end by noting that: "This weaving together of science and culture to communicate the relevance of science to everyday life seems to lie behind the growing popularity of these festivals."

Non-Traditional Venues

Similar to the examples above about dinosaurs in airports or physics at football games, a vast number of other places exist that can be co-opted for informal learning when the public is not expecting such an experience. In my class I refer to these as "non-traditional venues," which can be found in the most unexpected places. Take, for example, the Smithsonian exhibit of the giant extinct snake *Titanoboa* in Grand Central Station, New York City (Welsh, 2012; Fig. 13.10). Exhibits like this can alleviate the boredom while waiting for a train. Just think of the millions of people that pass through spaces like this every day. It is curious why

Fig. 13.9 The Cambridge Science Festival in 2015 (Cambridge Science Festival, 2018; photo courtesy of the University of Cambridge).

Fig. 13.10 *Titanoboa* exhibit at the Grand Central Station in New York (Smithsonian Channel photo).

more malls do not have STEM spaces; indeed, a precious few do, such as the aquarium in the Mall of America (2018) in Minnesota. It is becoming clear, however, that the demographics of shoppers are radically changed by online shopping. As such, the use of malls for non-traditional learning will likely become less important in the future.

A model program for informal learning in non-traditional venues is the Golden Gate Bridge. Since it opened in 1937, billions of cars have crossed this iconic structure (e.g., in fiscal year 2016–2017, 41 million cars crossed the bridge *every month*). As part of an informal learning agenda, the Golden Gate Bridge has a visitor welcome center with STEM attractions related to the bridge (Fig. 13.11). Despite the need to actively choose or plan to stop, this destination has more than 10 million

Fig. 13.11 Golden Gate Bridge Welcome Center (National Park Service, Kirke Wrench photo).

Fig. 13.12 Solar Walk in Gainesville (Howard L. Cohen photo).

visitors annually (Golden Gate Bridge, 2018), which surpasses the attendance at many large museums.

Closer to home, in 2002 the Alachua Astronomy Club (2018) installed monuments along one of Gainesville's major roadways representing the planets of our solar system. This so-called "Solar Walk" (although few people actually walk it) is a mile-long stretch representing the relative positions of the planets scaled to their positions in our solar system (Fig. 13.12). Similar to many roadway exhibits, most of the people riding in the vehicles that travel along this roadway do not stop (and there is no parking) to investigate further. Nevertheless, the Solar Walk has become an iconic installation within Gainesville's cultural attractions (Visit Gainesville, 2018).

In summary, the possibilities for informal STEM communicated in non-traditional venues is enormous. Learning researchers might quibble with the fact that, relative to K–12, little is known about what people actually learn from these opportunities, particularly when speeding along in a car at 50 miles (72 km) per hour. Nevertheless, other venues at which people can interact with (e.g., walk by) the display of STEM content are potentially more likely to engage in meaningful informal learning.

Non-Formal Science Education Programs

We have described the dichotomy between the notions of formal classroom (e.g., grades K–16) learning and informal free-choice learning, the latter of which typically includes unstructured activities such as a family strolling through a museum. In reality, however, a third category, non-formal science education (Blakey, 2015), includes structured activities oftentimes delivered within informal settings. These might include public museum lectures (Fig. 13.13) and also other science-themed groups (e.g., clubs), TED Talks, after-school STEM activities, and summer science

Fig. 13.13 Public events at Florida Museum of Natural History. Top: Evening science talk (FLMNH, Jeff Gage photo). Bottom: Outdoor science celebration at a local brew-pub (FLMNH, Kristen Grace photo).

classes for youth. Another general characteristic of these kinds of activities is that they have neither formal lesson plans aligned to curriculum and learning standards nor tests (assessments). They also are typically "free choice" (*sensu* Falk & Dierking, 2002).

Talks about STEM can also be presented at civic organizations like Rotary Clubs and senior living centers. Another movement in non-formal science education that has grown in popularity is the "science café" concept in which the audience will gather at a restaurant, have refreshments or a meal, the opportunity to socialize, and then enjoy a themed talk (Fig. 13.13). In recent years our museum has partnered with a local brew-pub to christen new beers and celebrate some natural history theme, like endangered butterfly species or the Tree of Life.

From a standpoint of Broader Impacts, non-formal presentations or activities have several positive benefits. In contrast to building an exhibit, they are relatively inexpensive and not as time-consuming to implement. They also have the potential to reach another kind of demographic that might choose to attend for another reason, such as the opportunity to partake in a new local beer. Museum educators are typically reluctant to schedule "one-off" talks without these being woven into the fabric of a larger theme, like an exhibit opening or a lecture series. It is therefore advisable to contact the potential organizers as soon as possible because many of them have themed programming organized months in advance. An important consideration of these kinds of activities is that their audiences can be quite diverse. In terms of reaching an intended audience, as we have discussed elsewhere, it is critical that the speaker optimizes the communication so that it will be accessible to them.

Traditional Mass Media

On Friday afternoons on my way home from work, I listen to *Science Friday* (2017) with Ira Flatow. He has a whimsical voice and the scientists that he talks with are interesting. Traditional media can include print (e.g., newspapers and magazines) and broadcast media, including radio (like *Science Friday*) and TV (either through news spots or longer shows, e.g., on public television). Similar to museums and non-traditional venues described above, traditional mass media has potentially enormous reach. A single media event or communication can reach a broad range of individuals across society, from thousands to millions of people. These kinds of media have their own idiosyncrasies, and in order to be effective, scientists need to understand the parameters. Most importantly, while some listeners might be predisposed to learn about science, in many cases the audience is the general public and therefore potentially heterogeneous. It is important for the scientist to communicate clearly without jargon. Although some forms of traditional mass media allow for more extended communication (e.g., nature series on public TV), others, such as news spots, require "sounds bites" and "elevator speeches."

Most of my experience with mass media has been with journalists and the press. A common concern about journalists is that, particularly for those that are inexperienced, they sometimes do not get the facts correct, or they incorrectly interpret what they have heard during interviews. This is not always the case, and these challenges typically occur with student reporters, particularly if they are inexperienced with science. In fairness to the profession, I have had many very positive experiences with journalists who have done an excellent job of communicating science to the general public. Although it cannot always happen (e.g., because of tight deadlines), it is best practice for journalists and science writers to allow their sources to read a draft of their press release or article. In so doing, the writer will receive valuable feedback and the opportunity to correct any factual errors before they are published.

This admittedly short description of traditional mass media only skims the surface. It also leaves out the immense and emerging impact of cyberenabled social media, which will be discussed in Chapter 15. Suffice it to say that traditional media has a potentially huge impact on informal learning and can also be repackaged for formal classroom learning.

Conclusions and Take-Home Messages

Informal STEM learning, or free-choice learning, can potentially occur in a wide variety of settings and throughout a person's lifetime (Falk & Dierking, 2002; Gewin, 2013). These settings can include built and other environments, some of which are intentional in terms of learning (e.g., museums), while others are unintentional, such as non-traditional venues. Informal STEM learning can reach different numbers of participants, either directly, such as through a science café, or millions through an exhibit or mass media. In all such cases it is important to know the intended audience and optimize the activity and style of communication to meet the learners' expectations. Each kind of these lifelong learning opportunities has its strengths and challenges – for example, building a museum exhibit sounds like a good idea, but it is among the most expensive and time-consuming of potential Broader Impacts plans. The advent of cyberenabled technology has revolutionized the dissemination of informal STEM learning, and any program or activity must be evaluated to understand its efficacy. It should also be noted that many of these activities simply cannot be done in a vacuum. They require a team of interested collaborators who span the gamut of content expertise, the learning sciences, and other STEM educational professionals such as exhibit designers and K–12 outreach specialists. Regardless of the specifics described in this chapter, Broader Impacts activities with informal, or non-formal, audiences can be both immensely rewarding and achieve significant benefit to society.

14 Public Participation and Community (Citizen) Science

School of Ants

During my Broader Impacts course, I heard about a new faculty member in UF's entomology department, Andrea Lucky, who was doing innovative community (citizen) science. I invited Andrea to give a guest lecture about her projects, including one called *School of Ants* (Lucky et al., 2014). I learned that using a simple yet carefully designed protocol, participants place crumbled cookies onto small cards on the ground, wait for a specified interval of time for ants to arrive, collect the ants, freeze them, and send their samples to Andrea (School of Ants, 2018). In her lab they are identified to discover the spread of invasive species of ants in the United States (Fig. 14.1). Her lecture in my class described the elegant simplicity of the project and its innovative public engagement. In particular, I was fascinated that, although lifelong learners of all ages are involved, many of the participants are elementary school kids (Fig. 14.1) and their teachers. Through an informative website, student participants learn about their contributions to Andrea's research. Few other community science projects that I know of have focused so effectively on harnessing the innate curiosity of kids to do science. Andrea's passion for science and public engagement represent the best of what scientists do in modern society.

Introduction

About 10 million people in the United States, or 5 percent of the working population, are employed in some aspect of STEM (Noonan, 2017). This number leaves more than a quarter of a billion other people in the United States who could potentially become engaged in the practice of science sometime during their lifetime. While it is unrealistic to expect full participation from the general public, any effort that increases engagement and broadens participation also promotes learning by doing, and thus science literacy. Public participation can also increase science identity for those who participate in the research enterprise. The general notion that STEM is primarily done by professionals – those who are paid to do so as their career – neglects the fact that few scientists were

School of Ants

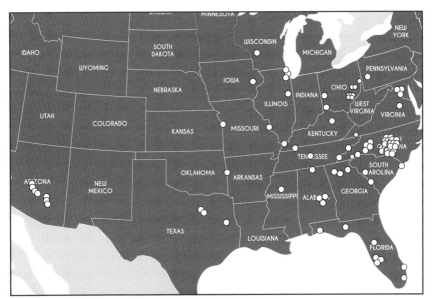

Fig. 14.1 *School of Ants* project. Top: Map showing localities of ants collected by participants in the project (redrawn from School of Ants, 2018). Bottom: Children collecting ants (Rob Nelson photo, with Andrea Lucky's permission).

professionals two centuries ago. Likewise, it neglects the more recent upsurge of interest among the general population to contribute to and engage with science. Public participation and "citizen" or community science (Box 14.1) are not new – for example, the National Audubon annual Christmas bird count started in 1900 (Cohn, 2008; Ellwood et al., 2018). In a modern context, public interest in science grew after World War II (Lewenstein, 1992). Over the past few decades more emphasis has been placed on active learning and engagement and thus termed "public participation in scientific research"

Box 14.1 "Citizen" versus "Community" Science

In recent years the term "citizen science" has been giving way to "community science," particularly in the United States. This relates to several factors, including the recent pejorative distinction between "official" citizens versus those members of societies who are not (e.g., recent undocumented immigrants or displaced people). In terms of social responsibility and inclusiveness, this artificial distinction is unjust. Everyone should have an opportunity to participate in science. The Audubon Society has a position statement as follows:

The word citizen was originally included in the term citizen science to distinguish amateur data collectors from professional scientists, not to describe the citizenship status of these volunteer observers. Today, however, it is important for us to recognize that the term has become limiting to our work and partnerships in some contexts.

Audubon (2018)

In addition, Audubon (2018) underscores the point that public participation is oftentimes a social or community activity, rather than a solitary pursuit. With this additional social context, the term "community" takes on even more relevance.

Therefore, where it makes sense in this chapter, the term "community" is preferred. The term "citizen" science will primarily be used in a historical context in which theory and practices are described, such as the seminal literature in this field (e.g., Bonney et al., 2009a).

(PPSR; Shirk et al., 2012). This participation also is linked to science literacy and the notion that a better informed public can make better decisions that affect not just their daily lives, but the overall progress of STEM.

Many aspects of science can be done by almost anyone (e.g., Budiansky, 1996), and in some domains potentially transcend educational, and to a lesser extent social, barriers. Participants from all walks and stages of life are typically labelled amateurs, avocationalists, and hobbyists, all terms with some baggage. For example, "amateur" can have a negative connotation, and the adjective "amateurish" can imply something done "in an unskillful and inept way" (Oxford Living Dictionaries, 2019). In a modern context, however, contributions from the public need not be amateurish. Instead, people making these discoveries can be thought of as engaged members of society who, along with professionals, advance knowledge in a meaningful way.

Public involvement in the scientific enterprise can come in several different forms and with different goals. These can include volunteers, community scientists, and crowdsourcing. Although in the past volunteers have sometimes been called citizen scientists, these terms are not synonymous. Volunteers participate for personal benefit, such as occupying leisure time and contributing to society. In contrast, community scientists are more actively involved in learning about and contributing to research. Crowdsourcing is a term borrowed from the business world that is related to large projects involving many participants, typically online. When learning is a goal, then these activities fall within the realm of informal STEM education. This chapter continues the thread of how people and communities learn outside of the formal classroom setting as they engage in public participation in scientific research.

Volunteers

I'm fully retired and want to give back to this country some of the things that it has given to me. I wouldn't volunteer if I didn't enjoy it.

Volunteer, Colonial Historical Park, Virginia (NPS, 1996)

The first category of public participation is volunteers. Some people in the education community disparage volunteerism because participant learning is neither an intentional outcome, nor is it carefully assessed in most instances. For this reason, volunteerism is not community science in the strict definition because, for the latter, expectations include active learning, data collection, and sometimes participation in the research analysis. Nevertheless, volunteers have an extraordinary impact on science through the work they do. For example, Syms (2016) reports that volunteers contribute more than a million hours of work each week in museums throughout the United States. In some smaller informal institutions, volunteers can outnumber paid staff (AAM, 2016). In many cases, volunteers working year-round involve retired citizens and their motivations are similar to the feelings expressed in the quote above. However, many non-profit institutions also value the contributions of younger volunteers who participate during their out-of-school free time. Studies have shown that best practices for volunteers include intentional training, mentoring, recognition, small perks (e.g., parking passes), efficacy analysis (evaluation), and record keeping. Thus, many institutions take volunteerism seriously and have one or more coordinators to train and keep track of this component of the workforce (AAMV, 2012). Large organizations, like the National Park Service (NPS), have capitalized on volunteerism and developed a national program of volunteers – VIPs, or Volunteers in Parks. These VIPs do a variety of work including physical maintenance, archiving documents on computers, and leading guided tours, to name just a few. Through the initial recruitment process, the VIP program seeks to align volunteer activity with the person's areas of experience and interest. Every year more than 250,000 people volunteer at more than 400 NPS sites (NPS, 2017).

Volunteerism does not have to be restricted to informal settings. During my sabbatical in the Santa Cruz, California, schools in 2015–2016, I observed many dedicated volunteers assisting teachers in the classrooms. Many of these had been teachers or scientists prior to retirement. Over the past several decades several strategic initiatives at the federal level have been geared towards promoting volunteers in formal education, particularly from the ranks of retirees (NRC, 1990). From a STEM volunteer workforce perspective, this practice makes lots of sense because the United States has more than one million scientists and engineers over the age of 60 who could be recruited to share their expertise in schools (Rea & Nielsen, 2010).

Volunteers at the Florida Museum of Natural History

In many respects our museum is an example of volunteerism within museums in the United States, and beyond. We have two general categories of volunteers: junior volunteers, aged 12–17, and adults. Over the past year about 100 junior volunteers logged 4,400 hours during summer activities, such as helping with youth programs. About 500 adults volunteered throughout the year and logged 34,400 hours of service. The adults participated in varied aspects of museum activities, including as docent explainers to the public (Fig. 14.2), as well as working behind the scenes in collections. Some of our dedicated volunteers have worked in the museum for 50 years. We have a full-time person to manage volunteer activities, and members of our staff in specific units (e.g., education or collections) are involved in supervising volunteers. The 34,400 hours worked over 2017–2018 roughly equate to the addition of 17 full-time staff working in our museum (FLMNH, 2018).

Within the research part of our museum, the vertebrate paleontology division (where I work) has had an ongoing volunteer program for half a century (MacFadden, 2017). We have greatly benefitted from the contributions of volunteers through their hard work. As an example, we have recently been excavating a five million-year-old fossil site called Montbrook, located in rural northern Florida. At Montbrook we are finding the remains of extinct elephant-like gomphotheres, horses, rhinos, and other extinct vertebrates that lived in what is now Florida during the late Miocene (Hulbert, 2016). From 2016 until mid-2018 we developed a large following of almost 700 volunteers at Montbrook, who collectively worked a total of about 12,000 hours excavating fossils. In addition to these statistics, about two dozen of our volunteers then spent hundreds of hours each preparing fossils in the laboratory and curating specimens into our collections as part of the overall effort related to Montbrook. Almost 500 volunteers and other interested people are, are members of a Facebook Montbrook Fossil Dig group where they share photos of field work and discoveries. Of the more than 25,000 specimens catalogued thus far in our research collection from Montbrook, about half were collected by the volunteers. Budiansky (1996) highlighted the fact that volunteers can make

Fig. 14.2 Volunteer (right) leading tour of the Florida Museum of Natural History exhibits (FLMNH, Kristen Grace photo).

surprising discoveries that advance the science of paleontology (Fig. 14.3). Hulbert et al. (2018) describes a skull of an extinct saber-toothed cat from Montbrook. This discovery, which was made by one of our volunteers – Bill Buhi (Fig. 14.3) – pushes back the origins of this rare group by a million years into the Miocene.

Supervolunteers

I recently heard an interesting talk about volunteers who help with the MAM (Mid-Atlantic Megalopolis) project (Skema, 2018; Fig. 14.4). The speaker talked about "supervolunteers," a cadre of a few highly committed volunteers who do the great proportion of the work and thus account for the majority of the productivity. Within any project or organization that involves volunteers, it is likely that several of them in aggregate do most of the work. This is also the case at our Montbrook fossil site, where a half-dozen of the most committed volunteers do most of the work. Thus, five of our volunteers worked more than 300 hours each; another dozen worked over 200 hours each either at the fossil site or back in the preparation laboratory and research collections.

Fig. 14.3 Dedicated volunteers at the Florida Museum of Natural History. Left: Bill and Carol Sewell working at the Montbrook fossil site. They have just finished making "plaster jackets" around fossil turtles that provide protection while transporting these specimens back into the lab (Bill and Carol Sewell photo). Right: Bill Buhi, a retired UF professor, working in the fossil preparation laboratory (Jonathan Bloch photo).

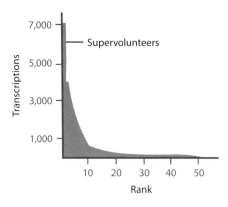

Fig. 14.4 The concept of supervolunteers, in which a few top performers transcribe the majority of museum data (taken from Skema, 2018). This model is generally correct in many instances – that is, a few volunteers (ranks 1–10) do most of the work, followed by a long tail of other volunteers who do smaller amounts of work, but in aggregate the latter are important to the overall effort.

Summary: Volunteers

As mentioned above, volunteers play an important role in the lives of cultural institutions and organizations such as, respectively, museums and national parks. The involvement of volunteers has occurred for centuries. While volunteers typically derive considerable benefit, such as giving back to society, and oftentimes volunteer for decades, this activity is not to be equated to community (citizen) science. The goals and expected outcomes of the latter are more geared toward active engagement, learning, and sustained participation in the process of science.

"Citizen" and Community Science

The terms citizen science, or citizen scientist, originated the mid-1990s (Bonney et al., 2009a). This movement has spread dramatically over the past two decades, with current involvement in thousands of projects worldwide. In contrast to

---

Box 14.2 "Moonshots" and the Billion Oyster Project

We choose to go to the moon.

John F. Kennedy, speech presented at Rice University, 1962 (NASA, 2018a)

Big thinkers, whether it be in politics, business, or science, like to envision projects so large that they are potentially mind-boggling but, if achieved, represent extraordinary leaps forward for science, technology, and society. These are sometimes referred to as "moonshots," in reference to President Kennedy's speeches in the early 1960s in which he challenged the United States to put a man on the moon during that decade. Consequently, US science and technology was mobilized through the Apollo space program and Kennedy's vision was achieved in 1969 (NASA, 2018b)

Box 14.2 (cont.)

In my opinion, the Billion Oyster Project is a moonshot. Oysters are a keystone species that are important to the health of shallow marine ecosystems. They also have significant economic impact on the shellfish industry. Two hundred years ago, New York Bay was the leading producer of oysters. Since that time, however, increased human activity and overharvesting have resulted in, respectively, increased pollution and a dramatic decline in the oyster populations in New York Bay. Launched in 2014, the Billion Oyster Project (BOP) is a partnership involving about 20 stakeholders from education, local restaurants, corporations, and non-profit foundations. It primarily engages thousands of middle- and high-school students from over 100 schools in all five boroughs of New York. The students recycle oyster shells from restaurants, use these to develop oyster banks for larvae, and monitor the water quality and development of local populations in the waters near their schools. The BOP seeks to restore sustainable oyster reefs to this region by the planting of one billion oysters by 2035; so far 25 million oysters have repopulated the New York Bay. In so doing, however, the BOP is providing a place-based learning opportunity for local K–12 students (BOP, 2018; Waterfront Alliance, 2014). At the same time, the BOP is contributing to real-life problems that affect humanity. The BOP has attracted the attention of the media as a model project that has elements of volunteerism, citizen science, and K–12 classroom learning (PBS, 2017).

(Agata Paniatowski photo)

volunteerism, citizen science weaves active participation and STEM learning into the fabric of the scientific enterprise. A related outcome is increased science literacy. Falk (2001) postulated that people will be more inclined to participate in citizen science if the research being done is relevant to their lives. Thus, the Billion Oyster Project (Box 14.2) has relevance to school-aged participants in New York. The citizen science movement has been enabled by the web and the building of online learning networks, also called communities of practice (CoPs; Lave & Wenger, 1991; Wenger et al., 2002). Citizen science projects can be described as a progression from less active to more active and sustained involvement (Inset 14.1; Shirk et al., 2012), and follow a logical progression (Inset 14.2; Citizen Science, 2018). As such, the public can contribute data or discoveries, collaborate with the scientists, and even co-create the research to be undertaken. Related to the latter, sometimes community participants are unencumbered with the minutia that frequently occupy scientists' minds. Thus, they are sometimes able to ask interesting questions that can both advance the research and be of broad significance.

Citizen science models

1. **Contributory** projects are generally designed by scientists, and members of the public primarily contribute data.

2. **Collaborative** projects are generally designed by scientists. Members of the public contribute data, but also help to refine project design, analyze data, and/or disseminate findings.

3. **Co-created** projects are designed by scientists and members of the public working together. At least some of the public participants are actively involved in most or all aspects of the research process.

Inset 14.1

Citizen science: Step by step

1. Scope your problem
2. Design a project
3. Build a community
4. Manage your data
5. Sustain and improve

Inset 14.2

Participants

Lifelong learners. The rise of citizen, and now community, science since the mid-1990s has attracted lifelong learners, oftentimes with a prior interest in science, such as amateurs and hobbyists. A study of citizen scientists in the United Kingdom found that only a small percentage of the potential population participate, with some bias toward white middle-aged males. A primary reason for participation is altruistic, that is wanting to help others to do science (Cambridge, 2016). In the United States, Jones et al. (2017) studied a sample of amateur astronomers and ornithologists. Similar to the study in the United Kingdom, Jones et al. (2017) found that the mean age of their survey sample was middle age and they were predominantly white; while the birders were of roughly equal gender, the astronomers were predominantly male. These participants spend about seven hours per week involved in their hobby or other leisure pursuit. Motivations for sustained engagement in leisure science include an internal drive and innate interest, the opportunity to socialize with other like-minded people, and increased science knowledge. Although lifelong learners do not necessarily equate with community scientists, many of the former can also contribute to science knowledge.

It is clear that the participation of lifelong learners in informal science does not reflect the demographics of the US population. Reasons for the lack of diversity within the lifelong learning community are complex. These likely relate to factors such as the disconnect between research objectives and community priorities within underserved minority groups (e.g., Pandya, 2012) and the amount of leisure time available. While there are exceptions to this rule, an area for opportunity for broadening participation in STEM is within these underserved communities.

> Citizen science opportunities provide real-life experiences that link what students are doing in school to what is happening in the world.
>
> Froschauer (2018)

K–12 formal settings. Although community science is oftentimes done in informal settings and associated with lifelong learners, many examples exist in which these activities are embedded within K–12 formal education. The World of Ants (2018) and Billion Oyster Project (2018) described above are good examples in this regard. In addition to the relevance of the activity to the students' world, these science projects also have many additional benefits such as working with real data and contributing to a larger and meaningful societal initiative. In a key resource for K–12 community science, SciStarter (2018) indicates that best practices for successful youth engagement have the students (1) taking ownership of data quality; (2) sharing their findings; and (3) engaging in a broader social community.

Froschauer (2018) advocates that citizen science can be started early in formal education (i.e., in elementary school). This begs the questions of whether these young scientists have the requisite knowledge to be able to participate, if the data that they collect are meaningful, or whether they are just playing at citizen science (Bourne, in SciStarter, 2018). The issue of data quality collected by non-professional scientists is not unique to young participants and is a common concern in most community science projects. Some kinds of data will be easier for young participants to collect, like ants, whereas others, such as taking measurements or identifying species, will be either more challenging or unrealistic in terms of data quality.

Crowdsourcing

Crowdsourcing

The practice of obtaining information or input into a task or project by enlisting the services of a large number of people, either paid or unpaid, typically via the internet.

Inset 14.3

Another type of community science that involves large numbers of people can be done via crowdsourcing (Inset 14.3; Oxford Living Dictionaries, 2019), which has expanded rapidly with the advent of computer-based projects available online. For many of these projects the participants

How it works

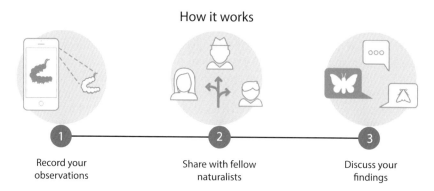

| 1 | 2 | 3 |

Record your   Share with fellow   Discuss your
observations   naturalists   findings

Fig. 14.5 The simple process of uploading species observations (iNaturalist, 2018).

are distributed geographically, oftentimes with an international reach.

iNaturalist (2018) is an example of a crowdsourced citizen science platform. People upload their observations about species occurrences, oftentimes using an app on a mobile smart phone. The occurrence data then undergo quality control – that is, they are curated. If these data are determined to be of research grade, they are then made available for aggregation by the Global Biodiversity Information Facility (GBIF, 2018). The statistics on participation and data generated are an impressive testimony to the power of crowdsourced science. More than 800,000 participants have signed up with iNaturalist. Collectively they have made more than 10 million observations of 167,000 species from around the world. Within the iNaturalist platform people can participate in specific projects, such as investigating the effect of the total solar eclipse of 2017 on life, or contribute to the overall data-gathering effort (Fig. 14.5). The platform is also designed so that participants can explore data, including an easily visualized distribution map, and if social interaction is a goal of participation, meet other members of the iNaturalist online community.

Related to online crowdsourcing, sometimes events can be concentrated into "blitzes," or intense episodes of work, like hackathon-style participation. The WeDigBio (2018) project promotes the transcription of old, mostly hand- or typewritten museum specimen labels into digitized format. Over the past several years, WeDigBio has sponsored four-day transcription blitzes that include groups physically transcribing from a single site (Fig. 14.6), or individual transcribers working online. With proper promotion via the web and social media, the increased and concentrated participation can yield significant productivity. In the four-day transcription event in 2015, more than 50,000 labels were transcribed from more than 20 sites involving thousands of individuals (Ellwood et al., 2018). These concentrated events therefore have the potential to do work, but also engage participants remotely using online platforms coordinated by initiatives such as WeDigBio.

The list of crowdsourced projects is expansive. In addition to iNaturalist, others such as Zooniverse (2018) or eBird (Box 14.3), to name only two, likewise have

Fig. 14.6 Participants in the WeDigBio 2017 transcription event at the Natural History Museum of Los Angeles County, California (Kelsey Bailey photo).

Box 14.3 Cornell Laboratory of Ornithology

No discussion of the history of citizen (community) science or crowdsourcing would be complete without highlighting the Laboratory of Ornithology at Cornell University. Founded in 1915 to primarily focus on research and graduate education, the Cornell Lab has grown to be a world leader in citizen science and public engagement. Its current mission is "to interpret and conserve the earth's biological diversity through research, education, and citizen science focused on birds" (Cornell Laboratory of Ornithology, 2018). With regard to active public participation, the Cornell Lab currently has more than 400,000 participants contributing to various projects related to ornithology, broadly construed to also include biodiversity, conservation, and related topics in biology and the environment. Fourteen million bird enthusiasts of all ages connect with the Cornell Lab online. Scientists and educators from the Cornell Lab have been leaders in the theory, practice, and evaluation of public participation and citizen science (Bonney et al., 2009a; 2009b; Dickinson & Bonney, 2012).

The community (citizen) science component has been sustained through a series of projects mostly funded by NSF, and have impacted a variety of lifelong learners, from serious bird-watchers to more casual backyard observers. In terms of crowdsourcing, the eBird (2018) project, in partnership with the National Audubon Society, contributes more than 7.5 million observations each month about bird species occurrences. An important outcome for these observations is the production of valid and useful scientific data. These crowdsourced data contributed are aggregated into large research databases such as the Global Biodiversity Information Facility (GBIF, 2018) and are used as the basis for scientific publications. The Cornell Lab has had more than 150 peer-reviewed papers based on data resulting from community science projects (Cornell Laboratory of Ornithology, 2018).

> **Box 14.3 (cont.)** 
>
>
> **Fig. 1** Birdsleuth Educator Workshop in the Great Smoky Mountains, a partnership between the Cornell Lab and Tremont Institute (Great Smoky Mountain Institute [2018], Tiffany Beachy photo).

large followings. Quality control is a common concern with data contributed by participants. How do these data contribute to the overall scientific enterprise in a meaningful way? Otherwise, if the data are not reliable, then although participants are learning how to do science, the data that they produce are of dubious research utility in the peer-reviewed world.

Can Community Participants Do Authentic Research?

> While scientists are often skeptical of the ability of unpaid volunteers to produce accurate datasets, a growing body of publications clearly shows that diverse types of citizen-science projects can produce data with accuracy equal to or surpassing that of professionals.
>
> Kosmala et al. (2016)

The process of doing science includes collecting data that become evidence to test hypotheses and make claims. As the quote above suggests, some practicing scientists are skeptical about whether the general public can meaningfully participate in science by contributing data. A large body of literature has been developed that addresses this concern (e.g., Cohn, 2008; Guerrini et al., 2018). The take-home message is that, with careful consideration, planning, and training, community scientists can contribute meaningful data in many, although not all, instances.

Data can take various forms, such as observations, identifications, collections, measurements, and discoveries. Observations can include bird counts, species can be taxonomically identified during bioblitzes, children can collect ants, new species of fossils can be discovered by the public, and participants can take measurements. But are these to be trusted? In terms of usefulness for the science, some of these are straightforward – for example, as long as the geological context can be confirmed, fossils discovered by amateur paleontologists represent important data that can be further studied by scientists. Likewise, as long as simple protocols are followed, children collecting ants should be considered of value. Observations and identifications can perhaps be more challenging, particularly when no voucher (i.e., a specimen deposited into a collection) is collected for subsequent verification. Nevertheless, many community science projects have protocols in which observations and identifications must be cross-validated by multiple participants. For example, iNaturalist (2018) requires at least three confirming observations of a particular species for the observation to be considered of research grade (Wiggins & He, 2016). Taking measurements is perhaps the most challenging expectation involving community scientists or crowdsourcing. Many scientists, including the author, prefer to take their own measurements, so this reluctance to use other people's data is not just a concern for "inexperienced" non-professionals. Within the context of taking measurements, some are straightforward, such as measuring the length of a bone, whereas others would be more challenging, if not impossible – for example, counting the number of pollen grains on a bee. Thus, not all measurements are of the same complexity and the community science effort should be aligned with the skills and experience necessary to collect meaningful data.

The age, background, and experience of the community participant can also affect what they are called upon to do. Many retired participants come from a STEM background, albeit mostly from other domains. From prior experience they understand the value of collecting valid data. Although skeptics may take pause, children can make meaningful contributions of data (e.g., collecting ants and fossils). In all cases the community scientist, regardless of age, will require training when it comes to more complex participation such as measurement. In some cases, for example fossils collected by school children, the data may be of some utility in actual research, but typically these activities are more effective in the students learning about how to do science (Conner et al., 2013). This is not always the case because some successful projects, like School of Ants (2018) and Dragonfly Detectives (Goforth, 2018), provide evidence that children in elementary and middle school can collect meaningful data.

In general, best practices for engaging the public in data collection will involve the researchers understanding the audience demographics, the kind of data that need to be collected to do the research (Costello et al., 2013), and optimizing the tasks that the community scientists can do relative to the effort involved in training. Outcomes have shown that the data collected from citizen participants are being used in peer-reviewed literature. GBIF (2015) cites 350 papers published in 2014 using biodiversity data, much of which is contributed by community scientists. Likewise, the Cornell Laboratory of Ornithology (2018) cites 150 papers based on data collected by community (citizen) scientists. Thus, if projects are properly designed and vetted, despite the skeptics, community science data can be, and are, used in peer-reviewed publications, and this trend will likely increase in the future.

Concluding Comments

> Public understanding and support of science and technology have never been more important, but also never more tenuous.
>
> Leshner (2012)

Public participation in STEM is of fundamental importance for science literacy in the twenty-first century. Science has generally been considered a trustworthy source of information that benefits society. Nevertheless, as the quote above implies, increased politicization, such as with regard to evolution and climate change, have polarized the public and cast doubt upon both the veracity and value of the scientific enterprise. Increased awareness, interest, and knowledge about STEM among the general public can lead to positive outcomes for society. One of the best ways to enlighten the public is through their direct participation in STEM. In a similar vein, some scientists have called for more of a two-way participation dialog with the public (Simon, 2010; Sykes, 2007).

Public participants can come from all walks of life, from K–12 education to lifelong learners. These can include volunteers, who traditionally are recruited to do work, and community scientists, the latter of whom engage at some level in the scientific process. The demographics of public participation have traditionally been skewed toward older whites. These stereotypes, however, are changing with increased emphasis on programs that seek to broaden representation. Similarly, the demographics are shifting as younger learners are participating in science through technology, via cyberenabled projects and platforms, for example. While the web has revolutionized public participation in science, it can also be isolating and negatively affect engagement. Indeed, many volunteers and participants expect or enjoy social interaction with the scientists and other participants. Scientists wanting to make meaningful inroads into broadening participation should keep these demographic trends and factors in mind.

Many public participants either enter a project with extensive prior knowledge, or gain knowledge during the individual projects. The ranks of participatory science such as astronomy or ornithology, to name only two areas, are full of serious and passionate amateurs. As described above, the negative connotation of amateur, or amateurish, is indeed unfortunate. These participants should be valued for this experience rather than treated as second-class citizens. In the world of paleontology, I have experienced numerous amateur paleontologists whose only difference in knowledge is that I am paid as a paleontologist, and they have other kinds of day jobs. I have noticed repeatedly throughout my career that amateurs are oftentimes better at finding fossils than the professionals, including me. I recall the time when three professionals, including myself, walked on the trail past a 33 million-year-old giant pig-like mammal in the Nebraska badlands, only to be upstaged by an experienced amateur who followed us and discovered the exquisite skull that we had not noticed.

Engaging volunteers and community scientists in a research project requires work and effort on the part of the scientists. These are termed "opportunity costs" (Ellwood et al., 2018). Successful programs do not just happen without care and feeding. The most effective kinds of programs will typically require coordination, training, and ongoing mentoring throughout the participants' experience. These kinds of projects also benefit from intentional stakeholder engagement, in a way that is sensitive to local perspectives and cultures (Lavery, 2018). Many public participants also expect to learn about research project outcomes and be part of a community; the latter typically are now enabled online. Likewise, volunteers want to be recognized and appreciated for their accomplishments, which can be profound. We have discussed the potential pitfalls and cautions with using community science data. Despite the skeptical professionals who might feel threatened, if carefully planned, scaffolded, and validated, many types of data collected by community scientists can significantly contribute to the overall research results.

A common concern of any program involving public participants relates to understanding if it was successful. For example, did it achieve the research goals as well as involve the public in a meaningful way? These questions fall within the realm of process of evaluation. A large body of literature exists about best practices to do evaluation, including many with a focus on community science projects (e.g., Jordan et al., 2012). It should be cautioned that this process is a science in itself and should not be taken lightly by scientists who think that by doing surveys they can demonstrate efficacy. In Chapter 18 we will discuss evaluation as this process pertains to many different kinds of Broader Impacts projects and activities. Suffice it to conclude here that public participation and community engagement in science and STEM is a big deal with immense potential for strategic impact in the twenty-first century.

15 Computers and Cyberimpacts

The NSF Project that "Keeps on Giving"

In 1996 I received supplemental funding for my existing research project to develop a website *Fossil Horses in Cyberspace* (FHC, 2017). The internet was in its early days, with people connecting to it via dial-up modems, which my students find as familiar as slide rules, 2 × 2-inch "slides," or typewriters. There was also no Web 2.0 (O'Reilly, 2005) interactivity like we have today. Over the past two decades, FHC has been used as a reference for K–12 projects, and we sometimes receive requests to reproduce photos from the website. What I did not realize until recently, however, is that FHC remains one of the most visited web parts of our museum's website (which contains more than 20,000 pages). For example, during the first quarter of 2018, the FHC home page was visited more than 90,000 times. Our analytics indicate that visitation spikes on certain days, with localized activity that seems to indicate schools using FHC for classroom learning. As also described in Chapter 7 (Fig. 7.1), FHC is one of the top returns for search engines with the query "fossil horses." All of this activity and longevity is despite the fact that, until recently, we have neither actively managed nor upgraded FHC at all. (My graduate students, most of whom were not yet born or were toddlers when FHC was launched, make fun of the term cyberspace, which they say reached its peak near the end of the last century.) FHC (2017) is a testimony to the potential and sustained reach of one particular website that far exceeded our initial expectations. Despite being created a quarter-century ago, FHC exemplifies the value of NSF-funded Broader Impacts using the internet.

Introduction

Twenty years ago when we used the prefix "cyber" in *Fossil Horses in Cyberspace* (see the opening anecdote), it was an innovative use of the term. Today it is ubiquitous in our culture. Over the past several decades we have entered the computer era and this technology has revolutionized our lives. Of relevance here, it also has revolutionized the ways in which we can do Broader Impacts to reach out for societal impact. Many of the activities described in this book have been affected, or enabled, by computer technology.

The National Science Foundation's (NSF) Broader Impacts merit review criterion was officially introduced in 1997 (Rothenberg, 2010). This coincided with the beginning of the widespread use of the internet by scientists. At that time many of us considered that an appropriate use of this new technology was to create a website about our research. Since then, however, and with the advent of Web 2.0 interactivity (O'Reilly, 2005), the internet has become so much more versatile. At the same time the bar has been raised and more is expected of an innovative use of cybertechnology in an effective Broader Impacts plan. This chapter will focus on cyberimpacts using several examples and case studies involving computers and the internet for communication, development of learning networks, the multiplying effect of the web, and social media.

Cyberlearning

Computers have revolutionized our ability to teach and learn over the internet. We can now deliver blended (a combination of face-to-face, online, synchronous, online and anytime; Maxwell, 2016) learning in a way that promotes access and breaks down geographic barriers.

In 2009, during the preparation of our Panama PIRE Project (Panama PIRE, 2017) proposal for consideration by NSF, the solicitation encouraged the use of cyberenabled instructional activities. I had never used videoconferencing before in teaching, but the thought that a course, or weekly project meetings, could be delivered to students and interns simultaneously at the University of Florida (UF) and Panama made lots of sense. We first had to determine the best way to connect. At that time, it was through Polycom technology and proprietary software. While the quality of the transmission was quite good, the limitations of the platform required considerable expenses approaching $10,000 per site for the video and audio teleconferencing infrastructure. A decade later the Polycom technology has become obsolete for us and we now use different platforms depending upon the need, including Skype, Adobe Connect, and more recently, Zoom.

For five years (2013–2017) I taught a summer course entitled "Informal STEM Practice" to preservice teachers (i.e., undergraduate STEM majors at UF). Half the class typically lived in Gainesville and came to the physical classroom, and the other half lived throughout Florida (and some beyond); all did internships in their local area. Without access to the videoconferencing platform, teaching this course with students both in class and connecting from multiple sites would have been impossible. As can be seen from the body language in Fig. 15.1, keeping all participants engaged is a challenge. This is even more so with mixed in-person and remote

Fig. 15.1 Students in the physical classroom and connecting online during the Informal STEM Practice course, 2012 (author photo). It is a challenge to keep the online participants engaged. The instructor needs to intentionally keep them involved, e.g., by asking questions, rather than having the virtual participants feel like outside observers.

participants. If not deliberately managed, e.g., through active engagement by asking questions of the remote participants (LaBorie, 2015), then engagement and learning will not be effective for everyone.

During other NSF projects, we have done K–12 outreach via classroom and role model visits. Although this can be arranged in person for local schools, when we do classroom visits in California, for example, using cyberlearning platforms is both time- and cost-effective. These are done mostly by Skype because the technological requirements are simpler using this platform and the point-to-point communication works well. The teachers typically frame the scientist's visit around relevant lesson plans. In order to promote student engagement, the students can ask questions directly of the scientist. We have also used videoconferences to prepare teachers living across the United States for upcoming field work with us. These are typically recorded and then teachers who could not attend the real-time meeting can connect remotely afterwards.

Webinars, or cyberenabled "classes" that teach or allow dialog from a presenter and online participants, are a component of many of our NSF projects. Similar to what we have done for teachers, we have used an hour-long webinar to provide context for amateur paleontologists, for example,

Fig. 15.2 Emily Graslie of the Brainscoop (2018) project, where she presented an hour-long webinar that had about 100 real-time participants. We recorded this webinar and made it available online; about 3,000 people had viewed it by the end of 2018 (Sheheryar Ahsan photo © Field Museum 2016).

helping us collect fossils in North Carolina. This webinar presented the background science and research objectives, as well as facilitating questions and discussion. It also provided the opportunity for people to meet one another online and therefore begin to build a community. With regard to attendance at these webinars, we offer certificates of completion, which are sometimes proudly framed and displayed in participants' homes.

Although cyberlearning can increase access and break down geographic barriers, our experience has shown that challenges remain because these platforms can stymie even the best of intentions. The computer technology and internet connection being used will frequently cause problems for an effective online learning environment. Simple things like using headphones and working from a quiet home environment rather than at a coffee shop with lots of ambient noise tend to lessen distractions and result in a better learning experience for the participants. Likewise, although some platforms like Skype have simple learning curves, other more complex platforms can be a challenging barrier to those who lack confidence in the use of these kinds of technologies. We typically ask novice participants to obtain headphones or the ubiquitous ear buds and do a trial run of our videoconferencing program before the actual webinar in order to minimize crises when multiple participants are connecting at the beginning of the webinar. Although not every platform has recording capabilities, those that do have the potential to increase reach. Our more popular webinar presentations have had upwards of 100 participants in real time (e.g., one by Emily Graslie, Fig. 15.2); the recorded sessions of some of these have been viewed thousands of times. These modalities and platforms therefore have the potential to increase the reach of one's Broader Impacts plan. In some domains, however, webinars may have reached saturation and are no longer considered a novel technology. Not unlike other communication, it therefore behooves the presenter to understand their audience and to present relevant and engaging content.

Development of Learning and Social Networks

Learning research has a theoretical framework to understand how people develop learning communities. Lave and Wenger (1991; see also Wenger et al., 2002) proposed the concept of a "community of practice" (CoP) to describe these entities, which include people working or learning together within a content domain as they share and develop common knowledge, practices, and approaches. Lave and Wenger (1991) observed CoPs among Yucatan midwives, Liberian tailors, navy quartermasters, and meat cutters, as did Brown and Duguid (2000) in the classic study of Xerox repairmen. Communities of practice have spread to many other domains in which knowledge is created and shared among members of the community. Certain attributes are required for a CoP to be successful, including: (1) a content domain of knowledge; (2) a community of interested people (i.e., the social fabric for learning); and (3) a shared practice within this content domain (Wenger et al., 2002). Levels of activity in a CoP can include: (1) a core group; (2) an active group; and (3) peripheral participants. Although CoPs were initially developed in face-to-face environments, they have expanded to include the building, maintenance, and improvement of online learning communities as well. Any scientist wanting to develop an online community to promote their research is well-served to understand the fundamentals of CoPs.

Online CoPs are widespread in the formal education community. These are demonstrated to be an effective means of communications for educators in domain-specific knowledge or for developing practice, for example (US Department of Education, 2018). In 2012 we started a project taking teachers to Panama with us, where they could gain authentic research experience collecting fossils. The teachers spent two weeks in the field with us in July, and then returned to their homes, which were mostly in Florida and California, to write lesson plans incorporating what they learned. As such, it was impossible for all of us, including the scientists and teachers, to meet in the same place. We developed the plan that during the fall semester scientists and teachers would participate in an online CoP via a monthly check-in hour. During this time, we would discuss progress, and the scientists were available to answer content-related questions. To be honest, the online component of this CoP was a dismal failure. Our summative evaluation (Davey, 2017) indicated that factors such as trying to find an hour within teachers' already busy schedules, particularly given that they lived in different time zones, was a challenge. Likewise, many of the teachers simply did not make the progress on the lesson planning that we had expected. In fact, many of them were planning to implement these lesson plans in the spring, so during the fall their attention was elsewhere. Despite the best intentions and planning, our idea to develop an online CoP that would continue scientist–teacher engagement in our project simply did not work. In subsequent years, we decided to scrap the online CoP idea and never found a perfect substitute with which to build a sustained online community. Other more successful strategies for

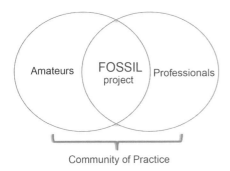

Fig. 15.3 Theoretical model of a CoP that brings together amateur and professional paleontologists (MacFadden et al., 2016).

continued engagement of the CoP included K–12 classroom visits, either in person or online, and getting together at annual cohort wrap-ups, or at national professional meetings.

Cyber CoPs are also pervasive in many other kinds of online informal STEM learning communities. Although some of these may not have been referred to as such, large online projects that develop knowledge and practice around a shared interest, such as ornithology (e.g., eBird, 2018), have many of the basic attributes of successful CoPs. Several years ago we sought to develop a national FOSSIL (2017) network to bring together amateur and professional paleontologists and their organizations, such as fossil clubs and museums. The theoretical CoP framework informed the process of bringing together potential partners and stakeholders – amateurs and professionals – via a shared interest in fossils and the science of paleontology (Crippen et al., 2017; Fig. 15.3). As is also described in this chapter, this project has developed a complex online and social network of almost 10,000 participants, mostly in the United States.

Social Media

Communicating with the public through social media is time-consuming, and most scientists are not trained for it.

Galetti and Costa-Pereira (2017)

To think that I would be writing anything at all about social media is quite astonishing. In 2013 I did not have a single social media account. At the urging of colleagues and students, I started to use Facebook and Twitter. That summer, I sent tweets once a day during my three weeks working with teachers in Panama. After several months, however, I lost interest in actively using social media on a sustained basis. It is well known that prominent scientists like Neil deGrasse Tyson have millions of Twitter followers. Although at a much lower level of engagement, many other scientists are active on social media and they are to be admired for this dedication. For me, however, it is difficult to keep up with emails, much less contribute to Facebook, Twitter, Instagram, or YouTube. It is also scary to think what kinds of social media platforms have yet to be developed that will gain large followings and then require another slice of people's time.

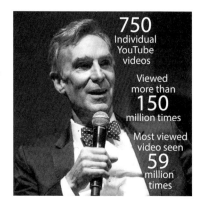

750
Individual
YouTube
videos

Viewed
more than
150
million times

Most viewed
video seen
59
million
times

Inset 15.1

Despite my skepticism, mostly stemming from lack of time, it is clear that social media as a form of science communication is limitless and here to stay. For example, many people are attracted to videos to learn about science and other forms of STEM, including those available on YouTube. One particular example is quite impressive: The highly visible promoter of public understanding of science, "Bill Nye the Science Guy," has more than 750 individual YouTube videos on a variety of topics that in total have been viewed more than 150 million times (Inset 15.1). The most highly viewed video, seen almost 59 million times, is a rap-themed simulated face-off between actors portraying Sir Isaac Newton and Bill Nye. Otherwise, many of Bill Nye's videos that discuss socially relevant topics including evolution versus creationism, climate change, and GM foods, each have millions of views (YouTube, 2018).

Case Study: The FOSSIL Project

It is clear that the value of social media to build online communities cannot be disputed. An example of this can be seen in how the FOSSIL project has developed since 2014 as a virtual community. When it started, we naively thought that our community would primarily be built through a listserv, our e-newsletter, and visitors to our website (FOSSIL, 2017). However, as can be seen in Fig. 15.4, we did not anticipate the contribution of social media to building our community since its inception. Before we launched our website, we relied on our listserv, newsletter, and limited use of Facebook and Twitter. The social media audience then expanded greatly in Facebook likes and Twitter followers (Lundgren & Crippen, 2017). The listserv never took off (it is now extinct) and the quarterly e-newsletter chugs along with modest growth relative to social media. The website continues to be a process of continuous refinement and upgrades. As evidenced by the number of members who have signed up to the FOSSIL website, it has experienced continued growth, although not at the overall rate seen by either Facebook or Twitter. In order to engage a potentially younger audience, in mid-2017 we started a FOSSIL Instagram account. With almost 1,500 Instagram participants as of the end of 2018, the rate of its growth in a year and a half exceeds what we have seen in Facebook and Twitter (Fig. 15.4).

Our findings so far indicate that social media participants constitute the majority of our networked community. Of the ~10,000 "members" of the

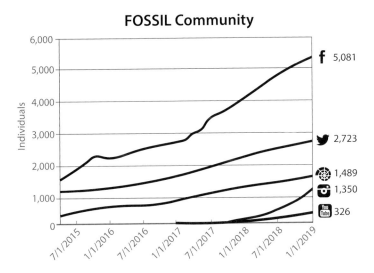

Fig. 15.4 Growth of the FOSSIL CoP learning network. This graph shows the increase in activity for the five ways in which the community interacts through online engagement: Facebook, Twitter, the FOSSIL website (spiral ammonite icon), Instagram, and YouTube. This is based on analytics and data gathered from 2015 until the end of 2018.

community, more than three-quarters connect via social media; only about 1,500 primarily connect by accessing the FOSSIL website, and fewer members access our e-newsletter (not shown in Fig. 15.4). Our analysis also shows little (< 5 percent) overlap of members active via social media also connecting to our website. Our original plan was to try to direct participation ("drive traffic") from social media to the website. This effort, however, was unsuccessful, so now we have stopped this strategy. We now realize that our community is not homogeneous, but is composed of participants coming from different entry points. Another lesson that we have learned is that social media growth does not just happen easily. In order to promote effective growth, the social media posts need to have certain kinds of content and be carefully coordinated (Lundgren & Crippen, 2017).

Mobile Phone Applications (Apps)

No discussion of cyberenabled technology would be complete without an understanding of the impact and potential reach of mobile smartphone apps. Over the past decade the growth of apps has been astounding. A Pew (2017) survey found that 77 percent of American adults have smartphones, a percentage that has more than doubled in just six years. In 2018, 3.8 million and 2.2 million apps were available, respectively, on Google Play and the Apple Store. In 2018, 205 billion apps were downloaded; this number is predicted to grow to 258 billion by 2022. In 2016, the global penetration of apps reached half of the world's population. Use of apps cuts across age and many social boundaries. Relative to other computer-enabled devices, apps have had a major impact on younger audiences. The average daily use of smartphones, tablets and similar devices with apps is about three hours for Millennials and two hours for Generation X, compared with less than an hour for Boomers (Statista, 2018). Even more so than computers, smartphone apps have democratized access to information and lessened the digital divide. In the United States, the percentage of Blacks and Hispanics that have smartphones is similar to Whites (Pew Research Center, 2017; demographic categories taken verbatim from source).

Of relevance to STEM learning, education is the third most popular category of apps and it has a large market in K–12. Although apps for K–12 learning have immense potential (Mills et al., 2017), issues remain about their use in the classroom (Metz, 2017). Of the millions of apps currently available, it is difficult to determine how many are science-related because of the way that apps are classified. Science content could therefore be found in numerous categories, such as education (as also for STEM), news, and travel, to name only a few. Several online resources describe apps for eBird (2018) and iNaturalist (2018; also see Chapter 14). Another science-focused app, NHM Alive (2018), enables a virtual tour led by Sir David Attenborough of the exhibits and research discoveries at London's Natural History Museum. If the user then opts to go to the museum, the app can be used to conduct a personalized tour. Access to news is another common use of apps; for example, science content can be reached via the *New York Times* (2018). The list of app resources continues to grow, with outlets such as Science Netlinks (2018) providing a clearinghouse to track these as they develop.

Hobbyists and amateur scientists are not fully engaging the next generation as has been done in the past, although online engagement has been increasing (Fox & Griffith, 2007). With this background in mind, it is clear that anyone wanting to use cybertechnology for Broader Impacts might well consider the potential of an app. Of relevance in my content domain, fossil clubs in the United States suffer from an aging demographic that threatens the future viability of their clubs. We also noted that our FOSSIL (2017) website did not fully engage the younger generation of those interested in science. We have therefore developed a myFOSSIL app (Fig. 15.5) that seeks to promote public participation. Modeled after successful apps such as iNaturalist, participants upload images of fossils collecting in the field to the "parent" website into a photo gallery. These are curated, and for those specimens that are of research grade (e.g., including appropriate metadata), our goal is to make these fossil specimens available for "Big Data" online research, again much the same way that eBird (2018) and iNaturalist (2018) observations are aggregated in GBIF (2018). One significant advantage of the mobile smartphone technology is the geolocation (latitude and longitude) tool. Location, or occurrence, data are fundamentally important to research-grade observations. Acquiring these kinds of data can be done within the app, unless a collection site needs to be kept private, in

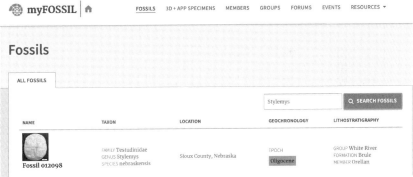

Fig. 15.5 Top: Screenshot of the myFOSSIL app with a photo of a 33 million-year-old (Oligocene) fossil tortoise from Nebraska. Bottom: The fossils uploaded to our online gallery at FOSSIL (2017).

which case the location function can be turned off. One of our graduate students in science education is currently studying the use of the myFOSSIL app as it relates to increased engagement and participation.

Thus, the development of an app is a potentially effective strategy to engage the public in STEM. Given the rapidly expanding field of options, however, it is important to understand what other existing apps already do. A unique niche can then be identified in the market for developing a new app. Other best practices include: (1) developing an app that serves both Android and iOS users; (2) developing the app for both smartphones and other mobile technology (tablets); (3) linking the app back to a website if data are being collected; and (4) unless you are experienced with the technology, employing the services of a professional app developer.

Virtual and Augmented Reality

> Digital nature: Are field trips a thing of the past?
>
> McCauley (2017)

The use of computers to simulate environments and immersive experiences is another area in which cybertechnology is rapidly developing. The potential applications for user experiences in VR or AR are limitless. In more sophisticated applications, the user is assisted by sensor headsets or eyeglasses (Box 15.1). Educational research is focused on how people learn using these devices, many of which are costly and not available in most classroom settings for formal education.

For the purposes here, we will focus on lower-tech alternatives in which virtual reality is simulated through apps, or a website, without the more complicated headsets or eyeglasses. Many educational institutions have developed virtual reality for geological field trips, for example. Some are "flatter," – that is, with maps and photos – and they lack the immersive feel. In contrast, some others employ 3D technology that allows the viewer to feel as if they are actually on-site. *Science Friday* (2016) highlighted a virtual field trip to the Sierra Nevada developed by Ryan Hollister, an educator in the Turlock Unified School District in California (also Derouin, 2018). In this virtual experience, participants go on a hike through the mountains and can even pick up and examine rocks to learn about them. Virtual geology field trips are not limited to education. The 3D Gaia platform uses virtual reality to help clients explore for oil and gas (Imaged Reality, 2017).

There are many other VR applications that could be used for Broader Impacts, such as 3D viewing of museum collections (also see Chapter 4) or virtual fossil digs. While it remains to be seen the extent to which the real environment can be fully simulated, there are many other positive attributes of VR. These primarily relate to access, for example, in classroom environments to allow remote locations to be toured virtually that would otherwise be impossible to visit. Another positive aspect of VR field trips is access for persons with disabilities who could not otherwise participate.

Box 15.1 Virtual and Augmented Reality

Fig. 1 Top: User experiencing virtual reality. Marxentlabs (2015, photo with permission). Bottom: Two young learners at the 2018 Indiana State Museum's annual GeoFest event, where an augmented reality app was used to view common Indiana fossils (Indiana Geological and Water Survey, Polly Root Sturgeon photo).

Box 15.1 (cont.)

With reference to McCauley's (2017) quote above (p. 204), it is unlikely that an immersive experience of a "real" field trip or holding a real specimen or artifact will ever be fully replaced. Nevertheless, new technologies such as virtual reality (VR) and augmented reality (AR) are revolutionizing access, particularly when the real experience or object are unavailable. VR places the user in an immersive experience, simulating reality. The computer uses sensors and algorithms to determine the orientation of the camera (Marxentlabs, 2015). In AR using an app, 3D images can "pop" on the computer screen. VR and AR are rapidly emerging as cyberenabled ways in which humans can interact with the world. No doubt we will continue to hear much more about these technologies in the future.

NSF's Grand Challenge of "Big Data"

In 2016 NSF issued a list of its grand challenges, followed by the "10 Big Ideas" for the twenty-first century (NSF, 2017p). This will guide strategic priorities for funding investments into the foreseeable future. While virtually all of the grand challenges are enhanced and supported by cyberinfrastructure, one specifically pertains to Big Data: "Harnessing Data for 21st Century Science and Engineering."

Many of our current scientific advances are being enabled by the data sciences. It is important that as discoveries are made within this domain, they are communicated to the public, such as through Broader Impacts activities. K–12 also needs to be engaged in this field. We have developed a pilot lesson plan for AP (advanced placement) biology in which students query big databases like the Global Biodiversity Information Facility (GBIF, 2018) or the Paleobiology Database (PBDB, 2018). The goal of this lesson is to understand ancient biodiversity patterns that address the misconception that horses (Family Equidae; Fig. 15.6) were introduced by the Spanish 500 years ago, when in fact the fossil record indicates that they evolved in North America (MacFadden, 1992).

Our experience has shown, however, that considerable scaffolding is required before high-school students can begin to query these biodiversity databases. First they need to understand the Linnaean hierarchy in order to search for the "Family Equidae." Then they need to understand how the database portal works as well as the back-end construction of databases and spreadsheet formats. In schools where this lesson was pilot-tested, computer literacy with even small database structure and searching is not part of the students' toolkit and thus big databases are too abstract. Some resources provide solid frameworks to scaffold understanding of databases (e.g., tes, 2018). A similar challenge for fossil databases (e.g., PBDB, 2018) is understanding the geological timescale. By scaffolding these basic concepts

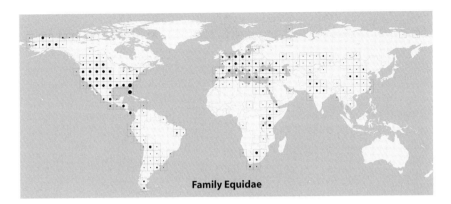

Fig. 15.6 Occurrence data for fossil specimens of the Family Equidae using GBIF (2018) over the past 55 million years. This map is based on 89,681 specimen records (with coordinates) and searching on scientific name "Equidae" and basis of record "Fossil specimen." The distribution shows the relative number of occurrences per box by the size of the filled circle (fossil horses never existed in Australia, for example). From this evidence one can conclude that fossil horses existed in North America in prehistoric times, contrary to the general public misconception about the origins of this group.

and tools, students can then query large databases in a similar way to the practice of scientists. There is no doubt that the use of Big Data for scientific research and education will become more popular in the future and provide twenty-first-century skills for the next generation.

Concluding Thoughts

The computer age has had a profound effect on societies worldwide. But for all of its benefits, cybertechnology has an emerging downside. Children in the United States now spend seven hours per day with various forms of technology. In contrast, they spend only 4–7 minutes playing outdoors (Kemple et al., 2016). This has resulted in a disconnect between the next generation and nature (Jacobson et al., 2015). In terms of social responsibility, scientists wanting to reach the public should understand both the benefits and pitfalls of cybertechnology for achieving Broader Impacts. In so doing, this understanding can potentially facilitate a transition from the cyber- to the natural world.

Cyberenabled technology is thus potentially an effective means to achieve enormous societal benefit. In terms of metrics, if the target audience is chosen correctly, and the appropriate tools and platforms are optimized, then relatively large numbers of people can be reached. This is exemplified, for example, with visitors to our museum, in which we receive more than 200,000 visitors per year "through the doors," but during the same interval we reach at least an order of magnitude greater number of "cybervisitors" to our website and via social media (Fig. 15.7). In a similar

Box 15.2 Supercomputers and "the Cloud"

Supercomputers consist of large "arrays" of computers and servers working together to solve complex problems and save and store large quantities of data otherwise not possible using smaller-capacity cyberinfrastructure systems. For the most part, supercomputer arrays are kept "behind closed doors" and therefore are not accessible to the public. If the public could see these installations, they likely would have a better understanding of the physical reality of, for example, "the cloud."

While public outreach is typically not an emphasis in supercomputer facilities, perhaps it should be. One exemplar is TACC, the Texas Advanced Computing Center at the University of Texas in Austin (Fig. 1, below). They have a staff devoted to public understanding, education, and outreach. I know about this first-hand, because in 2018 I took a guided tour of TACC and it opened up the physical world of supercomputers to me.

Fig. 1 High-school students touring the "Machine Room" at the Texas Advanced Computing Center, University of Texas at Austin (TACC photo).

way, our FOSSIL project, which includes Facebook, Twitter, Instagram, and You Tube (Fig. 15.4), comprises almost 90 percent of our online community. With millions of followers and views on, respectively, Twitter and YouTube, the reach of prominent scientists like Neil deGrasse Tyson and Bill Nye is undeniable. You will

Florida Museum 2017–2018

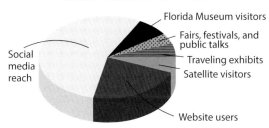

Fig. 15.7 The multiplying effect of cyberimpacts (FLMNH, unpublished data, 2018).

Social media (68%) and website (26%) account for 94% of the activity.

notice that the term "reach" is used, which is a relatively easy metric to determine and it also implies access. Neither of these, however, show a measure of impact, which implies a more deliberate process of determining both learning and affective responses. In most cases, impact typically requires a framework of assessment and evaluation (also see Chapter 18).

The old adage "if you build it, they will come" does not necessarily apply for outreach activities. With the burgeoning ways in which people can occupy their time, competition for cognitive engagement is keen. Likewise, many audiences have preferred means of accessing information. The dream of having a social media post go viral is just that, mostly a dream. Whether it is social media, or another form of twenty-first-century communication, and ultimately impact, in most cases one needs to develop a deliberate plan and understand best practices (Halford, 2016). The days are gone when one can develop a website or publish a Facebook post, and these activities constitute part of a social media plan. The bar has been raised and scientists need to consider cyberenabled means to engage and communicate in innovative ways for the benefit of society.

16 Developing a Broader Impacts Plan

The Last-Minute Request

From 1997 to 2004 I was in charge of exhibits and public programs at our museum. One day I received a phone message from a colleague in another department that went something like:

> Hi Bruce. I need to submit an NSF proposal tomorrow afternoon before I leave for international field work. I hear that you do Broader Impacts in the museum – perhaps we could do an exhibit together. I wonder if you could send me some boilerplate to use in my proposal? It would have to be no more than a page because my Project Description is already 14 pages long and only 15 pages are allowed. You can just email it to me. Although I'm busy getting ready to go into the field, if we need to talk over the phone, I'm available tomorrow morning. Let me know, thanks.

I called him the next morning and explained that I did not have any boilerplate text for Broader Impacts. He then asked if we might want to partner to do a small exhibit. Once we discussed what an exhibit might be like, I explained what it would cost. He then said that because his budget was tight, he did not have any funds that could be allocated towards Broader Impacts and the exhibit. I told him that there was nothing that I could do for him. That was the end of the communication. I do not know if he ever got his project funded.

Introduction

The Broader Impacts plan is fundamentally important to a successful NSF proposal. It can make or break the project. In the opening anecdote, there are at least four flaws in how the principal investigator (PI) went about developing his Broader Impacts plan; he: (1) had no clue about his audience; (2) waited until the last minute to start thinking about what he wanted to do; (3) left about one page for the Broader Impacts boilerplate text at the end of his Project Description; and (4) did not plan to budget any funds for these activities.

In this chapter we will first address these deficiencies and then discuss some other aspects of developing an effective Broader Impacts plan, one that will help with the goal of developing a successful NSF proposal. Although most PIs focus on the merits of the proposed research, in some cases an innovative, well-crafted, and integrated Broader Impacts plan can actually "carry" the project. An example of this is our Panama PIRE (2017) project. The core research of this project was to study ancient biodiversity in the New World tropics based on fossils collected along the Panama Canal. Overall the reviewers and panel ranked the research as solid. Nevertheless, they identified some risks and concerns that seemed to prevent the research from being judged at the highest level, including, respectively: (1) what would happen if we did not find any fossils; and (2) the research goals could have benefitted from more clarity. In contrast, our Broader Impacts plan, which proposed several elements, including the development of a bilingual exhibit that hundreds of thousands of people would see at the new Biomuseo (2018) in Panama, received glowing praise. I interpret, therefore, that the Broader Impacts were judged to be stronger than the proposed research. Large projects like the PIRE (NSF, 2009a) program are highly competitive; less than 5 percent of proposals submitted were funded. I believe that for our Panama PIRE Project, the Broader Impacts "carried" the research, and resulted in funding.

Know Your Audience

It is doubtful that the PI in the opening anecdote had given this topic much thought. We will focus here on how the audience relates to the development of a Broader Impacts plan. This includes: (1) the composition of the proposal review panel and NSF program officer; and (2) the audience that you intend to reach with your plan.

The first hurdle is, of course, getting the proposal funded. Not all panels are the same because they are typically composed of different individuals, so one cannot know specifically who they will be. The program officer will populate the panel with reviewers from different backgrounds and points of view, depending upon the goals of the program. Traditionally, a proposal submitted to a core program in a research (R&RA) directorate has been reviewed by a panel of scientists within the content domain. In recent years, however, even this practice has changed so that many panels will have reviewers representing Broader Impacts, diversity, and, depending upon the size of the project, evaluation. A proposal submitted to the education directorate will have more panelists representing learning, diversity, and evaluation, with less emphasis on the STEM content. Proposals submitted to this directorate will therefore need an increased focus in these fields and ways of thinking. Proposals submitted to very large programs (e.g., PIRE) might even have a panel member who is less familiar with the specific research, but will evaluate other aspects like project management and sustainability.

The second consideration is the intended, or target, audience for the Broader Impacts plan. For example, graduate students, lifelong learners, K–12 teachers and students, and rural underserved audiences can vary in their demographics, knowledge, and interests, and attempts to reach these groups must consider these differences. Likewise, there may be audiences with whom a PI will never communicate directly, such as mass or social media, and museum or website visitors. In all such cases, the Broader Impacts plan will need to optimize the engagement of the intended audience segment. An effective Broader Impacts plan will describe as fully as possible the demographics, educational level, and perceived interests of the audience. Developing a newsletter intended for teenagers will be far less effective than dissemination via an app or Instagram. The old adage "if you build it, they will come" only pertains to outreach that is carefully researched and planned, and optimized for the intended target audience.

Crisis Broader Impacts Development

Preparing a well-organized NSF proposal is difficult enough and should not wait until the last minute. This strategy just increases the stress associated with the process. I typically steer away from colleagues who work in crisis mode. Likewise, the sponsored projects or grants office at many institutions require several days' lead time to review and approve the proposal; sometimes it is sent back to the PI for further work. Our sponsored projects office recently instituted a hefty fine for proposals submitted to them less than 48 hours before the official NSF submission deadline. This practice is becoming more commonplace with proposal processing offices at research universities in an attempt to discourage last-minute submissions that thrust the office into crisis mode. Sometimes delays and snags are inevitable, but the more that one can manage the proposal development process in a deliberate way, the lower the stress level for everyone involved.

Some of the same recommendations are also true for developing a Broader Impacts plan. Instead of waiting until after the research has been planned and written, a better strategy is to start thinking about Broader Impacts at the beginning of the proposal development stage when the research questions are being formulated. This advanced planning will oftentimes result in a plan that is more effectively integrated into the project from the start. There is another important reason why Broader Impacts should be started early in the proposal development process: Most Broader Impacts require partners (e.g., K–12 schools or museums) and consultants (e.g., technology and evaluation). Using an example from K–12, it is ludicrous to think that a PI will go into a school and do the teacher a favor by providing STEM content for their classes. In any of these kinds of situations, unless relevant partnerships already exist, it takes time, effort, and trust to develop these connections. Likewise, once the partnership is established, if additional institutions, businesses, and budgets are involved, this will add additional layers of complexity and time to project development, such as with subcontracts.

Thus, in order to develop meaningful collaborations, an effective Broader Impacts plan should be started as early in the proposal development process as possible.

Having said this, it is also true that one can go it alone and focus on developing one's own website or engaging with one's own students. However, these will likely be perceived (by reviewers) as less innovative and effective relative to the Broader Impacts plans presented in other proposals to the review panel.

Broader Impacts Boilerplate

According to the Policies & Procedures Guidelines, NSF (2016b) requires a separate paragraph in the Proposal Summary and section in the Proposal Description that explicitly describe the Broader Impacts plan. In fact, proposals submitted without these specific sections are not in compliance and will be returned without review – typically after the deadline has passed, so one must wait another year to do a better job. While there are common elements of a successful Broader Impacts plan, most of these are unique to the proposal content scope, solicitation-specific requirement, and intended target audience(s). I have heard of other PIs and grants offices developing boilerplate text or templates, but I never use them – nor is it recommended. In the best-case scenario, the Broader Impacts plan is woven into the relevant sections of the Project Description, and then summarized in the required section near the end of the proposal. For example, when describing the graduate training and research participants within the Project Description, the PI can communicate a strategy for recruiting underserved minorities as part of their lab group. This strategy embeds within the proposal a Broadening Participation component that promotes diversity, but is described before the specific section on Broader Impacts. The one-size-fits-all practice of isolated boilerplate narrative that can be cut-and-pasted into the space available on the last page of the Project Description is simply a recipe for mediocrity.

Budget for Broader Impacts

In the introductory anecdote, the PI pushed back on the notion that Broader Impacts activities would require a budget. We all understand that budgets are tight, but even in NSF's research programs, a budget line item specifically for Broader Impacts activities is typically expected. This allocation would be particularly justifiable in the case of collaborations with external partners such as museums and schools. In these cases, the partner is less likely to become involved unless there are some resources brought into the collaboration. When new partners are invited into a project, it is best to think about "mutual benefit," or what each will get out of the partnership. A collaboration in which one partner derives positive benefit and the others receive little or no benefit will not last. In the best-case scenarios, projects in which mutual benefit is well developed will exemplify the adage "the sum is greater than the individual parts."

How much should be included for Broader Impacts in the budget? This will vary with the specific program and intent of the solicitation. The National Science Foundation provides little guidance in this regard, primarily because it seems like they do not want to be too proscriptive. For well-conceived and innovative Broader Impacts activities, a budget line item of a few percent – up to 10 percent of direct costs (i.e., excluding indirect costs [overhead]) – is reasonable. Some PIs believe that reviewers of research proposals will push back if these budget line items are too large, but some reviewers look at the budget to see if the Broader Impacts activities would be funded at an appropriate level. I have also heard the comment from unenlightened PIs that they simply cannot put money into the budget for these activities because the budget is already too large. This is a red flag indicating that the PI does not value the importance of Broader Impacts as part of their project.

In many instances, if the proposal properly integrates and identifies the Broader Impacts activities during proposal development, then these can become part of the overall plan. If, as mentioned above, the project explicitly will recruit underserved minorities – for example, as part of an undergraduate or graduate student research assistant plan – then these funds are a contribution to the overall Broader Impacts plan within the proposal. Although these may not be a separate line item within the Broader Impacts part of the budget, they should be discussed in two places within the Project Description, first under the student recruitment section and then briefly summarized again in the Broader Impacts section.

Another approach is to consider the Broader Impacts budget as seed money that will allow these activities to get started at a modest level. As these become successful, they can then be used as pilot data in support of a larger proposal submitted at a later date – one that leverages the main grant (Box 16.1). This was done successfully in the Panama PIRE (2017) project, in which we leveraged the Broader Impacts outreach plan first through an NSF supplement and then through a successful "stand-alone" RET (Research Experiences for Teachers) project (GABI RET, 2017).

In summary, a comprehensive Broader Impacts plan should be integrated early during the proposal development phase. It can include activities embedded within other parts of the project (e.g., support for underrepresented minority students), as well as budget line items to develop collaborative programs and activities. The budget for Broader Impacts, even in a research proposal, can be an actual allocation – albeit one of modest proportions – within the budget. Some PIs say that they had to cut out the budget line item for these activities when the award was being recommended at a reduced funding level. From the author's point of view, if the reviewers reacted favorably to the proposal, particularly if they mention the value of the Broader Impacts plan, then to cut this activity out after the review process is unethical.

What are NSF's Broader Impacts?

Although the core value of societal impact ("prosperity" and "general welfare") has been inherent in NSF since its founding in 1950 (Fig. 2.1), in 1997 Broader Impacts were formalized as one of the two (the other is Intellectual Merit) review criteria for all proposals submitted to NSF (Rothenberg, 2010). Ever since that time, PIs have struggled to understand what Broader Impacts are in the eyes of the review panel and program officers, the latter of whom make the funding recommendations. A common thread of these activities has been societal impact. Some

Box 16.1 Leveraging NSF Broader Impacts Activities

Fig. 1 Opening day of the *Megalodon* exhibition at the Florida Museum of Natural History, 2007 (FLMNH, Eric Zamora photo).

In 2004 I received funding from the GEO Directorate for a research study of the evolution of body size in large Cenozoic sharks, primarily *Carcharocles megalodon*, or "Megalodon." In that project, a small Broader Impacts budget line item was included for outreach – presenting talks to fossil clubs – but the proposal also mentioned that additional funds would be requested through a separate proposal to develop a traveling exhibition. For the latter, I collaborated with our head of exhibits, Darcie MacMahon, and we submitted an ISE (Informal Science Education) supplement to capitalize on the development of *"Megalodon: The largest shark that ever lived."*

We received funding for the *Megalodon* exhibition, which opened at the Florida Museum of Natural History (Fig. 1, above). Through many fossil specimens and interactive displays, it tells the story of the evolution and biogeography of *C. megalodon* and related sharks based on the funded research. In addition to the physical exhibition, *Megalodon* is supported by a website (FLMNH, 2018) and teacher resources. This has been, by far, the most popular of all of the traveling exhibitions that we produced "in-house." Since opening in 2007, it has traveled to more than 20 museums, science centers, aquaria, and other informal science institutions in the United States and has been viewed by about 1.5 million visitors (Fig. 2, below). The cost of marketing and maintaining the *Megalodon* exhibition has been sustained during this time via participation fees paid by the host institutions.

Box 16.1 (cont.)

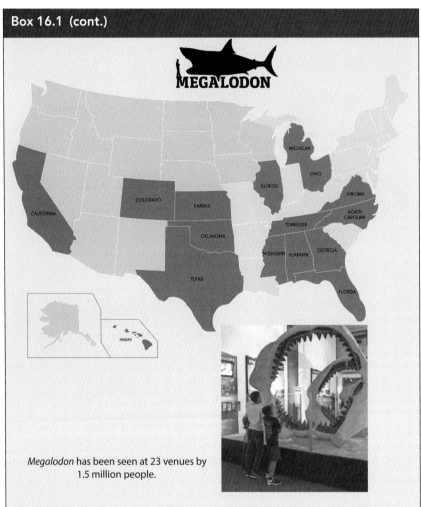

Megalodon has been seen at 23 venues by 1.5 million people.

Fig. 2 *Megalodon* visitors, including a map since this exhibition began to travel in 2007 (FLMNH, unpublished data).

PIs have equated this with outreach, which is frowned upon in some academic circles – that is, it is thought of as being less important than, and taking away from, research (Mervis, 2007).

In the most recent version of the annual proposal preparation guidelines, within the section on Merit Review, the Broader Impacts criterion is succinctly described as:

> the potential to benefit society and contribute to the achievement of specific, desired societal outcomes.

> (NSF, 2018h: III-2)

While this articulates a vision for what is expected, little additional guidance is provided on how to achieve these outcomes. Some previous versions of the annual proposal guidelines have provided more explanation, but similar to the terse definition of Intellectual Merit (NSF, 2018h), it is assumed that the PI will know how to do Broader Impacts. This assumption will grow to be correct some time in the future as the next generation of PIs will only know the NSF world with Broader Impacts. Nevertheless, I still receive comments from faculty knowing neither what is expected from them, nor how to effectively do Broader Impacts.

Although it is now more than a decade old, the Dear Colleague Letter (DCL; NSF, 2007), which was considered in the years following its publication to be too proscriptive, provides useful examples of categories of representative Broader Impacts activities. The more recent proposal guidelines (PAPPG; NSF, 2016b) provide a list of potential Broader Impact outcomes, many of which have their origins in the 2007 DCL (Box 2.1). The 2007 DCL provides five categories and kinds of Broader Impacts activities as follows:

1. Advance Discovery and Understanding while Promoting Teaching, Training, and Learning

Since the early days of the Broader Impacts review criterion, the default for many academics has been to propose they would incorporate the results of their research in the classes that they teach. Or that the NSF-funded research project would involve the training of graduate students. Both of these activities are now expected and a Broader Impacts plan with these components would likely be viewed as uninteresting. The bar has been raised and PIs wanting to incorporate a component of teaching, training, and learning into their plan would be well served to think about some innovative aspect(s) to make it more appealing. There are many new instructional and mentoring methods that could make such a plan interesting within the content of translating discovery and knowledge within higher education. If, however, the PI is not a professor at a college or university, then this component of a Broader Impacts plan would not be as obvious to include. As such, other avenues exist in this regard, including K–12 and informal STEM education.

Thus, the other large potential audience for this component is K–12 education. Blanket statements about intending to work within the K–12 environment to translate research to STEM teaching and the next generation of learners will be carefully scrutinized by reviewers, particularly if they have had experience within this audience or come from a background of learning research in formal education. Evidence of an understanding of STEM learning standards (e.g., NGSS, 2013), best practices for professional development, teacher and student demographics, planned evaluation, and a clear description of the K–12 partnerships already in place, will make for a more convincing Broader Impacts plan within this segment of society (Fig. 16.1).

Another way to communicate research discoveries is via informal learning environments. This might include, for example, museum docent training, public research lectures, and science cafés. Similar to some of the best practices described earlier for K–12 outreach, clear evidence of partnerships and evaluation will be important, as will an understanding of the audience to be served via these activities.

Fig. 16.1 *Fossils4Teachers!* professional development workshop held at the Florida Museum of Natural History in August 2017. Scientists, amateur paleontologists, and teachers came together to study fossils and develop lesson plans for K–12 classrooms (Pirlo et al., 2018; FLMNH, Jeff Gage photo).

2.   Broaden Participation of Underrepresented Groups

The concept of engaging a diverse segment of society in the science and technology (now STEM) workforce has been part of NSF's core values since Bush's (1945) report, *Science: The Endless Frontier*. This component has thus been important in developing Broader Impacts plans. It also seems that with the current NSF leadership – Director France Córdova – and the NSF-wide INCLUDES (NSF, 2017n) agenda, broadening participation is of even greater importance. I recommend that if one uses the 2007 DCL for the framework for developing a Broader Impacts plan, then PIs do not need to include something from each of the five elements. Likewise, there is nothing wrong with including multiple strategies within a single category, such as engaging diversity within one's research group as well as reaching out to underrepresented minorities in the general public. With the current culture of diversity and broadening participation within NSF, it is a good idea to include this component within the Broader Impacts plan.

The Broadening Participation plan is best developed within the context of a specific target audience or audiences. Reaching out to enrichment classes at a local K–12 magnet school is not compelling unless high percentages of underrepresented minorities are part of the student body (Fig. 16.2). It is far more compelling to include less well-funded and underserved (e.g., Title I) schools, teachers, and students in the Broader Impacts plan. Thus, if one wants to reach out to impact K–12 education, an analysis and description in the proposal of the demographics of the teachers, students, and schools to be involved is expected. It seems obvious, but underrepresented minorities can come from diverse

Fig. 16.2 Lisa Jones, elementary school teacher in Watkins Elementary School, Washington, DC (US Department of Education photo).

backgrounds and learning experiences. It is incumbent upon the proposer to demonstrate an understanding of these demographics. For example, in areas with high Latinx populations (e.g., in urban or rural underserved communities), an effective strategy might be to have graduate student role models visit classrooms and teach in Spanish, particularly in schools where English is a second language (ESL), or if the school is dual-immersion (English/Spanish). Likewise, a female African-American graduate student talking about careers to African-American K–12 students is far more effective than a stereotypical role model of an older scientist (like the author) going into the classroom.

3. Enhance Infrastructure for Research and Education

Certain institutions have the capacity to develop infrastructure and that can be shared with others. Natural history museums build research collections (Fig. 16.3) with sustainable resources that are available, hopefully, in perpetuity. These collections are thus made available for research by scientists. Larger research universities have specialized laboratories and major instruments that are typically made available to other researchers, such as from smaller teaching universities that lack these kinds of resources. More recently, the use of large centralized computational facilities, like supercomputers, frequently serve a broader community. Many of these kinds of research infrastructure have received prior or current NSF investments. As such, there is an expectation that they will be made available to larger audiences and stakeholders.

While it is easy (and somewhat expected) that this infrastructure be made available for research, the potentially more innovative plan is to make these resources available for education. A 3D-scanned file of digitized museum collections can be

Fig. 16.3 Behind-the-scenes in the McGuire Center lepidoptera research collections at the Florida Museum of Natural History (FLMNH, Kristen Grace photo).

used for K–12 education via 3D "prints" (Morphosource, 2018; oVert, 2017; Fig. 16.4), thus making specimens available that are rare, or originals that cannot be loaned for education. Likewise, summer professional development for teachers (e.g., CPET, 2018) can use actual laboratories at college and universities. Many large "supercomputers" are housed in facilities that are rarely seen by the public. Innovative public education and outreach programs, such as that developed by the Texas Advanced Computing Center (TACC, 2018; Box 15.2 Fig. 1) can clarify abstractions such as "the cloud" and enhance public understanding of STEM.

4. Broaden Dissemination to Enhance Scientific and Technological Understanding

A typical Broader Impacts plan will propose a dissemination strategy within one's own professional community. This will include presentations at national (and international) professional meetings, as well as publications in peer-reviewed journals. This plan, however, has been used too often and is no longer sufficient. Dissemination needs to be broader than the narrow confines of one's own academic or professional community.

As with most other dissemination strategies, the plan should start with an understanding of the intended audience to be reached. It should then be optimized so that the broad dissemination will be of interest and thus engage this segment of society. Proposing to develop a museum exhibit (Fig. 16.5) based on recent or ongoing research potentially can reach large audiences, depending upon the annual visitor numbers of the host institution. As we have discussed before, although exhibits are very time-consuming and expensive to build, they have the potential to reach large, informal museum audiences. Many other strategies exist that can reach the intended

Herpetology downloads

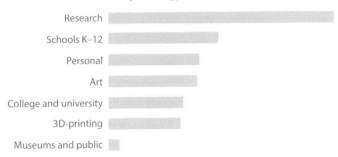

Research	
Schools K–12	
Personal	
Art	
College and university	
3D-printing	
Museums and public	

How downloads are used

- eighth-grade classroom instruction

- 3D print for surgery planning

- ...to make a 3D resconstruction of *Siren intermedia* with XROMM. This work will hopefully be published in a paper.

- Import into Houdini, do cool stuff, post on my Instagram

Fig. 16.4 Downloads from Morphosource (2018) of 3D-digitized specimens from the Herpetology Division at the Florida Museum of Natural History (UF). Specimens from the UF Herpetology Collection have been downloaded more than 7,500 times. Although about half of these downloads have been used for the original intended purpose – research – the graph at the top also indicates the other uses for these images. The lateral view of the Tuatara skull (*Sphenodon punctatus*, UF: herp: 11978) shown has been downloaded more than 175 times and used for a variety of purposes, as indicated by users' comments.

audiences, ranging from mass media (print and broadcast) to those enabled by the internet such as hosting a website, or a deliberate social media campaign.

5. Benefits to Society

This last representative activity in the 2007 DCL encompasses the overall vision for Broader Impacts. Societal benefit has been embodied in the culture of NSF since its founding in 1950, which has been "to promote the progress of science; to advance the national health, prosperity, and welfare; to secure the national defense; and for other purposes." Thus, in addition to basic research (although the human health agenda was subsequently moved to NIH), NSF seeks broad benefits to society.

We must remain mindful of the fact that NSF is funded by taxpayer monies, and at a fundamental level, NSF-funded research must be relevant and give back to society in a way that can be understood by the "person on the street." In addition to advances

Fig. 16.5 Temporary museum display of Miocene fossils collected from the Montbrook fossil site based on recent field excavations in northern Florida (Jeff Gage photo).

that improve life in the United States, such as computers and technology – like the now ubiquitous mobile phones – research funded by NSF must address issues of societal concern, such as public understanding of evolution, climate change, and genetically modified crops. Another societal benefit accrued from the outcomes of NSF-funded research is developing a more diverse STEM workforce in the twenty-first century.

Congressional oversight is a constant for funds that are appropriated for NSF. Scientists must do a better job communicating the importance of their research so that periodic and persistent challenges to the value of science are informed by the most recent advances. It is also incumbent on NSF and PIs to better inform decision-makers (e.g., in the US Congress) so that society can continue to benefit, either directly or indirectly, from scientific advances. Although somewhat more vague than the previous four representative activities described in the 2007 DCL, this activity is fundamentally important in the foundational justification for why we do NSF-funded science. Broader Impacts are essential for improving public understanding of, and continuing support for, government-funded science. Through Broader Impacts, the PIs are contributing back to society for its overall welfare.

I frequently turn to Bruce Alberts for wisdom about the importance of science in society. In 2010, Alberts received the Vannevar Bush Award for public service in science and technology. Harkening back to Bush's (1945) report, Inset 16.1 (in Pinol, 2010) exemplifies Albert's remarks while receiving this award.

"Science truly represents an 'endless frontier' through which we continuosly advance our understanding of the natural world around us..."
– Bruce Alberts

Inset 16.1

An important component of an effective Broader Impacts plan not explicitly described in the DCL (NSF, 2007) is the concept of how you will know whether you were successful. As the research discoveries and knowledge are being produced, it is important to know what the goals will be for the Broader Impacts plan. Some plans can result in deep impacts for fewer beneficiaries, such as novel mentoring or teacher professional development, although the latter can have indirect impacts on the students that the teachers teach (the "train-the-trainer" model). Other activities can impact large numbers of people, but it is unclear to what extent they are learning about STEM. Evaluation of the efficacy of the Broader Impacts plan is important to any project. It also follows that the larger the project scope, the greater are the expectations for documenting efficacy, or success with these activities. We will discuss this further in Chapter 18.

Summary Recommendation and Best Practices

The best scenario is to start planning Broader Impacts activities early in the proposal development process. This might also include, for example, engaging partners and stakeholders. If these activities require specific funds, an appropriate amount can be included in the budget. It is reasonable to propose a varied slate to Broader Impacts activities that relate to some, but not necessarily all, of the five representative activities described in this chapter. I usually recommend that the choice of these should be of special relevance to the project. For example, if one works in an urban setting, engage underserved minorities specific to the local demographics. If the local informal science institution (e.g., a museum) has a specific emphasis of relevance to the research, propose a strong linkage. If graduate students are engaged with social media, propose innovative uses for this kind of communication. Try not to fall into the trap in which PIs think that they must do something from each of the five categories (this seems to be one of the reasons why the NSF guidelines have become less specific or proscriptive). In most situations, quality is better than quantity. Choose several activities, particularly those that are not contrived, that uniquely relate to the project, and do them well.

17 Project Management and Sustainability

No More "Seat-of-the-Pants" Project Management

In 2010 we received notification from NSF that our Panama PIRE Project would be funded. Similar to other PIREs (NSF, 2009a), this was a large five-year international research and education project with lots of complexity. During the proposal approval phase, our NSF program officer let us know that all PIRE projects needed to develop a detailed management plan. I had never done one of these before, but I also had not previously received an award of this size, scope, and duration. We were asked to respond to questions that expanded upon some of the personnel, management, resources, and sustainability sections in the main proposal. In addition, however, our management plan, which ended up being 16 single-spaced pages with flow diagrams and a Gantt chart, also included discussions of SMART goals, Likert scales, meta-goals, communication plan (including PR and branding), "sparks," dispute resolution, risk management, an expanded evaluation discussion, and of relevance here, sustainability.

As we started to develop an initial draft of this plan, I wondered why NSF was requiring us to do this? At the time it seemed like yet another level of bureaucracy that we had to do in order to receive the funding. Nevertheless, we submitted the plan and it was apparently acceptable, because we did not have to make additional modifications. In retrospect, however, it made lots of sense. The National Science Foundation wanted to see that we understood and had made plans to manage such a complex project, and this document was a written guide for us. During the project we followed this management plan to a large degree, and even with changes to the project along the way, we referred to it on many occasions as the Panama PIRE (2017) project unfolded.

Introduction: NSF Investments

The National Science Foundation (NSF) has developed a culture that by funding projects they are making *investments* in research and education. Implicit in this notion is that NSF is a partner in the project. Thus, there is an added expectation of resources provided by the host institutions receiving the funding, as well as other stakeholders. A corollary of NSF's expectation is the notion of sustainability of projects – that is, that some components of the project will continue after NSF funding ends (Inset 17.1; Oxford Living Dictionaries, 2019). The noun "sustainability" and adjective "sustainable" have dramatically increased in frequency and public awareness over the past several decades (e. g., UNESCO, 2010). A decade ago the notion of sustainability, or sustainable projects, likewise became more commonplace at NSF, but initially within a different context, that of sustainable science. During this time programs were developed and projects funded that promoted environmental sustainability, such as those dealing with water and climate. Programs like this were packaged into a larger NSF-wide initiative called SEES (NSF, 2017o), Science, Engineering, and Education for Sustainability, and related "spin-off" solicitations, some of which are still active. In addition to the infusion of sustainability within the research culture of NSF, a legacy of this "movement" over the past decade is the additional expectation that projects funded by NSF are sustainable to some extent.

Definitions
Project management
The practice of initiating, planning, executing, controlling, and closing the work of a team to achieve specific goals and meet specific success at the specified time.
Sustainability
The ability to be maintained at a certain rate or level.
Example
"schemes to ensure the long-term sustainability of the project."

Inset 17.1

The processes of project management and sustainability are strongly linked. Without effective project management, plans for sustainability after NSF funding ends will have a lower probability of success. This chapter describes some of the essentials of project management that facilitate sustainability, and in so doing, maximize the benefit of NSF's investment in scientific research and education projects.

Project Management

In 2011 we were developing a complex, multi-year iDigBio (2018) project funded by NSF that would include the coordination of hundreds of institutions around the United States to digitize a few hundred million specimens in museum research collections. Given the scope of this project, we were advised by the NSF program officer to hire a professional project manager, perhaps with an MBA degree, and not to worry about whether the person had content expertise in digitized biodiversity collections and museums. We realized that it would be unlikely that we would find someone with expertise in both domains. Schwalbe (2016) describes the seven optimal skills of project managers, including that they: (1) are highly organized, (2) take charge and know how to lead, (3) are effective communicators, (4) know how and when to negotiate, (5) are detail oriented, (6) recognize and solve problems quickly, and (7) possess the necessary technical skills. Scientists possess some or many of these skills, which they typically learned as "soft skills" during their education. The difference, however, is that while scientists find many of these skills as an avenue to doing what they want to do – the science – project managers are focused on project management as the goal in itself, and the specific research content is secondary to their framework of goals and activities.

Returning to the example of iDigBio, we initially found a professional project manager from the business sector who did well, but left the project after one year (see the section on "risk" later in this chapter). We then found another person who had previously worked in industry. He was a professional project manager who has subsequently acquired sufficient science domain expertise to be highly effective. His role has become integral to the success of the planning and administration of the iDigBio project. He also now conducts project management seminars during our annual summit at which participants from other museums can benefit from his expertise and way of thinking. Without a full-time professional who has been primarily concerned about the management of iDigBio, our communication, planning, staffing, budgeting, reporting, and accountability would have been far less successful. While it is unlikely that all projects require this level of professional expertise, most funded by NSF can benefit from an understanding of project management and, in the context here, how this can support planning toward sustainability.

The corporate and business sectors understand the interdependency of sustainability and project management. Both of these can also be thought of as part of the strategic planning process within an organization (Martens & Carvalho, 2017; Silvius & Tharp, 2013; Silvius et al., 2012). In the for-profit business sector, effective project management and sustainability are embodied within the corporate goal of making money; poorly run businesses are not sustainable and either cease operations or reinvent themselves. Likewise, in the nonprofit sector, effective project management and sustainability plans can optimize the mission, goals, and strategic impacts of the organization. While the fields of project management and strategic planning are far too broad – nor does the author have the requisite knowledge to do justice within the limited scope here – best practices and recommendations do exist that can inform researchers about the sustainability of NSF projects.

Essential Elements of Project Sustainability

Effective project management includes a series of elements, including staffing, timing, scope, budget, and resource allocation (Schwalbe, 2016). Most projects are also finite and go through a process or life cycle (Fig. 17.1). In addition, the four elements listed below are fundamental to any effective project and will also be critical to planning toward sustainability after NSF's investment. All of these were included in the topics that we were expected to cover in our management plan for the Panama PIRE project described in the opening anecdote.

Communication

Nothing that we do in science (and most other walks of life) can be done without effective communication. Communication is woven into every aspect of projects.

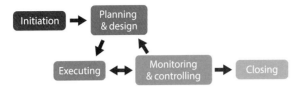

Fig. 17.1 The five components of the project management cycle (Alphamu57, 2018).

With regard to project management, communication is essential to the functioning and success of the project during every phase. The style and mode of communication is essential to the internal working of the project team, including the principal investigators (PIs), staff, and students, as well as partners, stakeholders, and advisors. In smaller projects, the modes of communication can be simpler and intuitive, and can evolve organically as needed. In more complex projects, including those with participants who are concentrated in one institution or area and others who primarily participate remotely, communication should be more carefully structured into a formalized plan that can become part of overall project management. In all cases, the style and mode of communication within project teams will most likely be a combination of structured meetings, frequent written messages via email, phone calls, and hopefully other impromptu communications. For remote project participants, email and videoconferencing may be the norm; it is, however, a best practice to have remote participants and those at the project hub periodically meet face-to-face – for example, if possible quarterly or no less frequently than at an "all-hands" annual project meeting ("summit").

I have developed some best practices with regard to project communication; none of these are particularly innovative, but they have worked well for me. Likewise, my style of communication has evolved over the decades, particularly with the advent of cybertechnology. Two of these are discussed here: email and remote communication. Many of us have been using email since the 1990s and it is both a blessing and a curse. Few people like to open their "Inbox" and see hundreds of new emails needing attention. I try to minimize the number of emails that I send to project participants, but there is no doubt that this is an easy mode of communication. Email works well for many everyday communications and also provides a written documentation of what was said. I have a few basic rules, however: (1) emails do not work well to answer complex questions, or for effective brainstorming; and (2) when a back-and-forth dialog becomes confused or adversarial, email no longer works well. For both of these, it is better to switch to other forms of communication; face-to-face meetings work best in these instances.

With regard to remote communication, face-to-face via videoconferencing, particularly during brainstorming, is preferred over a phone call and is certainly better than email. With regard to a hybrid meeting format, in which a group of project participants are together in a physical room, combined with others connecting remotely, videoconferencing is strongly preferred over the remote participants connecting via phone. While teaching classes in real-time with students in the physical classroom and others connecting remotely (e.g., Fig. 15.1), it is important that the remote participants are not passive observers of the dialog developing in the classroom. It is important for the leader to actively engage the remote participants because this communication typically does not happen organically. Understanding the challenges of engagement for remote participants, if I have an option of walking across campus for an important meeting in which I plan to actively participate, or connecting remotely, I typically opt to be physically in the room.

Strategic Planning

> In preparing for battle I have always found that plans are useless, but planning is indispensable.
>
> Dwight D. Eisenhower (Wikiquotes, 2018)

In the mid-1990s the chair of my academic department requested that I lead the faculty in developing a strategic plan to guide our programmatic vision for the future. At the time, I had no experience in this process, and wondered if it was just another bureaucratic waste of time. We went through a deliberative process of discussion and goal setting. Ultimately, our strategic plan was signed off by every faculty member, and it was filed away in some drawer to become part of our departmental archive. If it was referred to as our departmental roadmap, this was infrequent. Nevertheless, the process of faculty engagement, consensus-building, and shared vision was the most enduring outcome of the process.

Planning is integral to the process of project management and fundamental to the ultimate success of the project (Nokes, 2007; PMI, 2018; Schwalbe, 2016). This includes an articulation of project vision, mission, goals, objectives, and indicators of success. Planning should not be rigid, but instead keep open the possibility of seizing new opportunities that advance the project's mission and goals. Thus, effective planning is also integral to the development of sustainable components for the project, ones that will continue when NSF's investment ends.

Flexibility

What happens if, during project implementation ("Executing" in Fig. 17.1), an unanticipated discovery is made that potentially can improve the research but was not in the original plan? The value of serendipity should not be underestimated in project management because it frequently (almost always) occurs during the course of research. Even the best-prepared plans may need to be modified as the project unfolds. Projects that adhere too strictly to the original plan can preclude the potential to innovate as opportunities arise. Schwalbe (2016) describes the importance of agile projects, in which quick changes can be made based on iterative development and collaboration. In addition to projects being *flexible*, or *agile*, another term sometimes used is *nimble*. The National Science Foundation recognizes the importance of agile project management. As such, PIs have some leeway to be flexible and consider innovations that might improve project outcomes. Incremental changes in response to new opportunities can be described in the annual project report, or if they are deemed to be significant changes that might affect the scope ("scope creep"), NSF program officers can be consulted for advice.

Iterative Improvement

Fig. 17.1 shows a feedback loop in which, during the implementation phase, a project is iteratively informed by "Monitoring and Controlling." No project plan is perfect from the beginning, and, as we have already discussed, opportunities happen that warrant changes. In the program evaluation world (Chapter 18) this is termed "formative evaluation." Large projects funded by NSF typically have a rigorous mid-project review, called a "site visit." A review panel, similar in scope to the original review panel that recommended funding, is convened to review progress and make suggestions for improvement. Depending on how the project is developing, particularly if there are serious deficiencies identified by the panel, this can be a grueling two-day, face-to-face process. The project team rarely looks forward to the site visit; nevertheless, these reviews typically provide some degree of constructive feedback upon which the PIs can act during the remainder of the project. Having been through grueling site reviews myself, in retrospect they are one of the best practices for optimizing NSF's investment for successful project outcomes.

Risk Management

What happens if a critical member of the research team leaves the project? Or if the experiment fails, permits are not issued, or fossils are not found? All of these constitute examples of potential risk. Years ago an NSF program officer assigned to our iDigBio (2018) project asked us if we had developed a "risk register," which at the time I had never heard of, although it seemed obvious what it meant. Schwalbe (2016) describes a risk register as a list of possible risks and the ways in which they would be managed in the event that they occur (i.e., a risk is "realized"). Personnel can be replaced, but when other factors out of the project team's control happen, these represent risks to the project goals and, ultimately, success.

Another kind of risk is inherent in research projects, and it also relates to the previous section on flexibility. As new opportunities arise, they may be pursued, but there may also be a risk associated with the new direction. If so, then the potential risks need to be considered relative to the risk tolerance of the project team. Generally speaking, taking risks has the potential to result in greater rewards, such as certain kinds of research breakthroughs. The National Science Foundation realizes the importance of risk and reward, but the problem is that the review panel may not be so inclined and sometimes will feel more comfortable funding "safer" research projects. The risk tolerance of NSF reviewers can be an impediment to major breakthroughs and discoveries that might come from high-risk research.

Sustainable Project Design: Hypothetical Example

The Gantt chart in Fig. 17.2 (also see Inset 17.2; Tech Target, 2018) shows a hypothetical three-year NSF-funded project. Its mission is to advance research, education, and dissemination using natural history specimens collected during summers. The NSF solicitation has specific statements about describing project management and sustainability within the proposal description. The initial conditions assume that the project will primarily be capitalized by the NSF investment. The PIs' project effort and infrastructure are funded by the host institution. No significant funding is currently anticipated after the three-year funding period provided by NSF.

Fig. 17.2 shows that a project manager will be recruited early in Year 1. Likewise, graduate students will be recruited starting in Year 1 and continuing throughout the three-year funded project. The PIs, project manager, and students constitute the core project team. They will be primarily responsible for planning and implementation of the project during Years 1–3. The support of the project manager and students will end at the end of Year 3. Engagement of partners and stakeholders, representing either individuals or institutions, will occur throughout this project. It is anticipated that they may remain active in some of the sustainable components after NSF funding ends.

> ### Gantt Chart
>
> ...is a horizontal bar chart developed as a production control tool in 1917 by Henry L. Gantt, an American engineer and social scientist. Frequently used in project management, a Gantt chart provides a graphical illustration of a schedule that helps to plan, coordinate, and track specific tasks in a project.

Inset 17.2

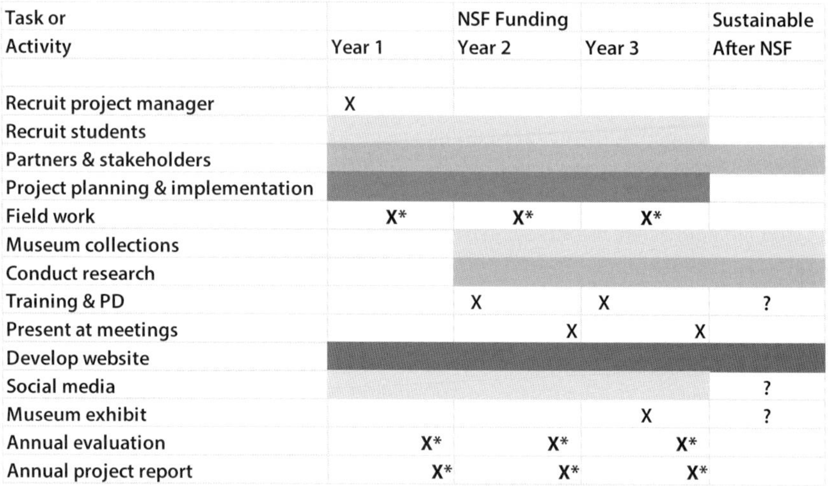

Task or Activity	Year 1	NSF Funding Year 2	Year 3	Sustainable After NSF
Recruit project manager	X			
Recruit students				
Partners & stakeholders				
Project planning & implementation				
Field work	X*	X*	X*	
Museum collections				
Conduct research				
Training & PD		X	X	?
Present at meetings		X	X	
Develop website				
Social media				?
Museum exhibit			X	?
Annual evaluation	X*	X*	X*	
Annual project report	X*	X*	X*	

Fig. 17.2 Gantt chart using a simple Excel spreadsheet format (also see Microsoft, 2018) showing tasks and timing of project components. Although sophisticated Gantt chart tools are available, this was done using a simple spreadsheet format. Three kinds of components are indicated: (1) "X*" representing a critical event or "milestone" that is required before other later activities can be undertaken; (2) "X" representing a specific event or time-limited activity; and (3) a horizontal bar, representing an ongoing activity. The position of the X* or X within the year roughly indicates when this milestone or other event will occur.

Field work to collect natural history specimens and associated data will occur during each summer of Years 1–3. These are critical events ("milestones") – the research and development of museum collections – which are core project elements, and are dependent upon the success of the field work, particularly during Year 1. Field work is a significant cost and is only planned during the three years of NSF funding. Museum collections will be curated by the graduate students during the project. After the project funding is over, these collections are available in perpetuity for additional study; these physical "products" (*sensu* NSF), as well as the research done on these collections, therefore represent a sustainable legacy of the project.

The project plan calls for a professional training and development (PD) workshop for participants external to the core project team during Years 2 and 3 using NSF funds. It is conceivable that if these workshops are successful and deemed of sufficient perceived value, then they could become sustainable, and the cost is then transitioned from NSF funding to the participants. However, this transition is best planned as an incremental shift of the costs from the project to its participants and stakeholders.

Dissemination of research results will take place at an annual professional meeting during Years 2 and 3. Given the cost, these are not planned to be sustained after NSF funding. A website and social media campaign will be developed and ongoing throughout Years 1–3. Both of these activities may represent sustainable activities after Year 3. A museum exhibit is planned for Year 3 to highlight the project and its research. Depending upon its design and audience, it might represent a sustainable project element after Year 3.

Annual program evaluation will be conducted during Years 1–3. This will include formative evaluation in Years 1 and 2 and summative evaluation in Year 3 (Frechtling et al., 2010; also see Chapter 18). The submission of the Annual Report near the end of each project year is dependent upon completion of the evaluation; both are considered milestones. A 90-day window is given by NSF at the end of each year, within which the report must be filed; most PIs do this in the final 30 days. It is not a good idea to wait until after the report deadline, because a series of computerized red flags come into play that affect not just the project in question, but cascade to block other NSF-funded projects involving the PI team. As a practical matter, if these red flags persist they can actually block additional funding increments until they are cleared.

To summarize, the Gantt chart (Fig. 17.2) has project elements that will end after NSF funding and project components that are identified, a priori, as sustainable. The latter include partners and stakeholder connections, which might lead to "spin-off" projects, the legacy of the museum collections and research done on them, and the website, and possibly social media – the last of these requires intentional activity unless some of the participants are particularly passionate about this form of communication. Other project elements that might be sustainable, if the cost is borne by participants and stakeholders, include the professional training (PD) and possibly the museum exhibit.

Other project elements not represented in the Gantt chart (Fig. 17.2) include such features as flexibility and risk analysis. What happens if an essential PI leaves the project? What happens if the field sites to collect the specimens are not available during the project? What will happen if a new technology is developed that would further enhance the research? As presented above, these are essential components of a project that must also be considered during planning and implementation.

To think that sustainable project elements will occur at no cost and additional effort is unrealistic. These elements will be in a maintenance phase, and not in further active development, but there are costs nonetheless. Thus, after Year 3 the research will still cost the proportional salary effort of the participants, the maintenance of the collections requires ongoing space and a collections manager, and the website requires ongoing IT infrastructure support. If the host institution or other partners are not able to provide any of these sustainable resources, then the project will indeed wither and die. This scenario is one that NSF might not consider to be a good investment.

Selected Best Practices and Recommendations for Sustainability

The natural cycle of a project is that it will end – the closing phase (Fig. 17.1). While NSF funds discrete projects, several mechanisms exist in which these can leverage and capitalize other projects in the future. A discrete NSF project will produce important knowledge and open up new and innovative avenues of future research and Broader Impacts activities as well. In this regard, partnerships are also important and provide potential to diversify future funding streams.

Pre-Award Planning

As PIs are developing their plan and proposal (Chapter 16), many of them do not push their home institutions about the resources that might be required to better support the project during NSF funding and afterwards. In the context here, this also can be a mechanism to ensure viable sustainability. A good time to leverage resources from the home institution is prior to the successful award recommendation from NSF. It takes more work at a time when PIs are already busy focused on getting the proposal ready for submission, but this effort is potentially time well invested. The justification for additional resources provided by the home institution has several benefits. If the project is funded by NSF, then it brings: (1) prestige to the institution that will presumably advance its mission (department chairs and deans like this); and (2) overhead funds that further support the institutional infrastructure (museum directors and research administrators like this). In the proposal, however, one needs to be careful about how such commitments are described. It is a delicate balance: on the one hand, some panel reviewers like to see institutional commitments, particularly if the PI is from a large, well-funded institution; on the other hand, some panelists may react negatively because of the added resources available to better-resourced PIs that may confer an unfair advantage in contrast to PIs from smaller, less well-funded institutions.

When we were developing the Panama PIRE (2017) proposal, we negotiated an annual contribution of direct funds from our vice president for research. These funds were immensely beneficial because most NSF projects have certain budgetary restrictions with regard to expenditures. The irony of the Panama PIRE was that it was supposed to encourage international cooperation, but because of general NSF rules, only US participants could be supported. For international partners from developing countries, it is unrealistic to expect significant funding for students and international exchanges. The funds provided by the vice president for research allowed us to fund these opportunities and thus strengthen the bilateral international training component of the project.

There are many other examples of successful leveraging of resources from the host institution. In many cases, university professors can negotiate a reduced teaching load during the project. As another example, a colleague in a state teaching university received new museum cabinets from her dean as a matching contribution for receiving an NSF award. These kinds of commitments can help during the NSF funding, like in the Panama PIRE (2017) project, or also support sustainable components after NSF funding – for example, the museum specimen cabinets. Many other stories like these can be told, and while some institutions are better prepared to help, most are interested in doing what they can to enhance the success of NSF proposals and research awards.

The Value of Partners and Stakeholders

Partners are collaborators who derive mutual benefit from involvement with a project. Stakeholders are others with a vested interest in, or who will be affected by, a project's outcome. Both of these kinds of collaborations are important to the success and long-term viability of projects like those funded by NSF. These can potentially add to the infrastructure as well as the diversity of participants and ideas that enhance the project activities.

Although most NSF programs encourage some collaboration for most of their funded projects, some – such as the PIREs – specifically expect highly developed partnerships. This is further explained in the NSF's solicitation 09-505:

> Proposers are strongly encouraged to develop synergistic collaborations with a wide range of U.S. academic institutions, including four-year and two-year institutions, in order to take advantage of complementary strengths in international research and/or education and broaden the impact of the proposed PIRE program. They are also encouraged to develop synergistic linkages with other domestic and international partners that will enhance the PIRE project, including research laboratories, professional societies, corporations, institutions of informal learning (e.g., museums, zoos), etc.
>
> (NSF 2009a)

The development of partnerships takes work and time to be nurtured. A simple "cold-call" email may start the communication, but meaningful (i.e., mutually beneficial) collaborations can rarely be effective at the last minute before the

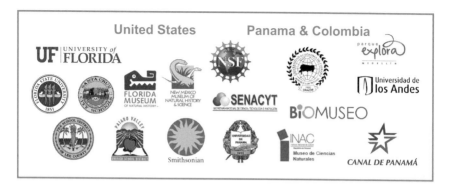

Fig. 17.3 Graphic list of partners and stakeholders in the Panama PIRE Project (Panama PIRE, 2017).

proposal deadline (such as shown in the introductory anecdote in Chapter 16). Our Panama PIRE (2017; Fig. 17.3) benefitted from a seed grant and then several years of intentional partnership development prior to NSF funding. Likewise, other partners were brought onto the project as it unfolded, and some became inactive as well.

Some of the most effective partnerships span different segments of society, or "boundary-crossing between systems" – such as industry and universities, which oftentimes can spark innovation (Kaufmann & Tödtling, 2000). Other common collaborations can include academic research and K–12 education that can promote Broader Impacts. If the partnership outcomes are of lasting importance, then when NSF funding ends other means of sustaining the project may be supported by the partners and stakeholders.

Renewals and Morphing into a New Project

In the strict sense, a renewal is a continuation of a prior research project funded by NSF. As the prior award is approaching the end of the project, a renewal request is resubmitted for a new increment of funding. There has even been a special category of proposals called accomplishment-based renewals (ABRs), although these are not recommended as a primary avenue of funding. Decades ago I submitted an ABR, which required a five (not 15) page proposal with up to five peer-reviewed publications attached, which were demonstration of accomplishment. The reviewers were supposed to evaluate the merits of the next funding increment based on past research performance. Despite the potential merits of this system, the reviewers were not accustomed to it and did not uniformly react positively. There was a TV commercial (about fast-food hamburgers) during that time; referring to this, one reviewer quipped "where's the beef?" in reference to the fact that the five-page Project Description provided insufficient information with which to evaluate the proposal. Given this experience, I have never done another one of these kinds of proposals again. Although NSF still considers ABR proposals (NSF, 2016b), I do not recommend them. In practice, renewals of any kind, whether they are ABRs or ones with more traditional 15-page formats, seem to be less frequent than decades ago. The concept of "more of the same," and incremental increase in knowledge goes against the NSF culture of innovative and transformational science.

Concluding Thoughts

For many PIs, developing a sustainability plan is a scary proposition that is frequently done during the later phases of a project. As advocated above, it is a good idea for the sustainable components of a project to be identified from the start (e.g., the hypothetical Gantt chart in Fig. 17.2). In so doing, the project management cycle can close down those components that will end, whereas the planning process can focus on ensuring the success of the sustainable components. In addition to sustainable components within the NSF-funded project, leveraging other resources to sustain project components is also a good idea. Additional strategies include renewal requests that build upon prior projects and morphing some components into other fundable projects, perhaps in different NSF programs (i.e., not strictly a renewal). Having a strong project plan and committed partners and stakeholders are prime prerequisites when looking for additional sources of funds after NSF's investment ends, thus ensuring the sustainability of certain aspects of the project.

18 Were You Successful? Evaluation and Metrics

The Evaluation Keynote Talk

In 2012 the PIs of all funded PIRE projects were invited to a summit at NSF. The program officer assigned to our Panama PIRE project asked me to present a keynote talk on project evaluation. Being surprised, I asked her: "Why me? I have no background in evaluation." She answered that our Panama PIRE project did an exemplary job of evaluation. When I told my evaluators back at my home institution that I was invited to a summit to present a keynote on project evaluation, they were likewise surprised. Two years earlier when our project began, it was a rocky road for me and the evaluators. I did not have a clue how the evaluation should be developed. Likewise, I begrudged the amount of time and money that would be spent on the annual project evaluations. I have since developed an appreciation of the importance of evaluation to fine-tune projects as they develop and, ultimately, to determine a project's success.

Introduction

Sooner or later, scientists with National Science Foundation support will experience some sort of evaluation, whether they like it or not. This is particularly true for larger projects with a greater emphasis on outcomes and impacts. The traditional notion in some principal investigators' (PI) minds, that funds are awarded for researchers to "do good things" without eventual accountability, is unrealistic in today's world. For a variety of reasons, the bar has been raised on being able to demonstrate the success of NSF projects, and in so doing, the value of the investment made using taxpayer funds. The problem with evaluation is that unless you are already involved in educational or psychological research, most STEM professionals do not understand what this process entails, how it is done, and how expensive it can be to do it right. There are also different kinds of evaluation procedures depending upon the project being analyzed. It also follows that, if done properly, evaluation is in itself a science with accepted protocols and best practices. Speaking from my own painful past experiences of trying to "do it on the fly," evaluation is best left to the professionals.

Definitions

Assessment versus Evaluation

In some circles, the terms "assessment" and "evaluation" are synonymous, but some experts make a distinction between these. According to Bell et al. (2009), *assessment* is the process of determining what and how individuals learn, whereas *evaluation* is the process of determining the effectiveness (also "efficacy") of a program, activity, or project. It is of course true that some programs can have an assessment component built into it, so these two terms are not always mutually exclusive.

Different kinds of learning have distinct goals, and thus can involve either assessment or evaluation, or both. In formal K–12 education the gold standard of any program, activity, or intervention is the impact on student achievement, which fundamentally starts with effective teachers (Rockoff, 2004). School superintendents are particularly interested in this metric when they account for investments made or programs undertaken. In contrast, however, assessing learning in informal STEM environments is more challenging, and has engendered a large body of theoretical literature that informs practice (e.g., Bell et al., 2009; Lloyd et al., 2012). Thus, a dichotomy has existed between understanding the efficacy of K–12 formal versus informal STEM learning. While the former can be quantified through, for example, pre- and post-"intervention" assessments, it is more difficult to determine the impact of an out-of-school experience on affective behaviors – that is, qualitative impacts on categories such as attitudes, interest, engagement, identity, skills, and behavior (Bell et al., 2009; Diamond et al., 2009; Fenichel & Schweingruber, 2010; Ferguson et al., 2014).

Evaluation versus Research

Although evaluation and research both collect and analyze data and prepare reports, each of these typically has different goals and objectives. Evaluation is geared toward providing an analysis and demonstration of the success of a specific project and, if it is still underway, how it can be improved. Evaluation typically is focused on the project without the intention to make the findings generalizable to larger audiences. The final reports produced during evaluation are intended for internal use, for example by the project team and funding agencies. Broader dissemination to the research community is typically not a goal of evaluation.

In contrast, the intent of formal learning research is to test hypotheses based on an established theoretical or conceptual framework. The ultimate goal of this research is to inform or advance a field of inquiry with generalizable conclusions, new insights, and recommendations for best practices. The data are typically collected using a systematic sampling methodology and carefully designed and validated instruments, and analyzed using sophisticated methods. The study can use qualitative data, quantitative data, or mixed methods (Diamond et al., 2009). Likewise, the sampling method can be quasi-experimental or via rigorous comparisons of treatment (via intervention) and control groups. The results are oftentimes intended for dissemination at professional meetings and publication in peer-reviewed journals.

Sometimes the dual roles of evaluation and research can collide within projects. Both of these processes involve data collection. Therefore, without effective communication among project participants it is sometimes unclear who is doing what. It is not a best practice to duplicate effort or develop similar surveys that are sent multiple times to participants or research subjects. It therefore is important for the project leadership team to understand the research goals as opposed to those of the evaluator and to plan accordingly. In the best-case scenario, with proper coordination, some data (e.g., demographics) can be collected in a single survey instrument that can be used for both evaluation and research.

How to Find an Evaluator and How Much Will They Charge?

Scientists new to the evaluation process may wonder how to find an evaluator. There are several factors to consider. In order to prevent a perceived conflict of interest, it is best to look for someone external to the project, particularly during summative evaluation (some NSF programs actually require this). Having an external evaluator is less critical for front-end and formative evaluation, and these can be members of the project team if evaluation expertise is represented. External evaluators can oftentimes be found in the education college at large universities, or by searching for a private evaluation firm (e.g., on the Center for Advancement of Informal Science Education website [CAISE, 2018]).

Individual evaluators typically have specialties. For example, an evaluator experienced in K–12 assessment may not know how to evaluate informal STEM settings, and vice versa. Larger multi-person firms can either specialize in formal or informal evaluation, or do both. It is also helpful if the person chosen has previous experience as an evaluator for NSF projects, if that is where proposals will be submitted. Likewise, if the project has, for example, a large online component or an app, then it would be best to find someone with prior experience with these kinds of dissemination tools.

Novices to the field of evaluation are oftentimes shocked by the cost of these services. Private evaluators can charge up to 10 percent of total direct costs for NSF-funded projects with a comprehensive and rigorous evaluation plan. Scientists do not appreciate that private evaluators oftentimes have considerable overhead costs (e.g., office space, utilities, computers, equipment, and software) that are covered in academic research institutions by indirect costs. Sometimes a graduate student or faculty member, oftentimes not associated with the project, can conduct evaluation for a stipend or summer salary; this can be far more cost-effective than going outside of the home institution. Similar to the budget line items frequently required to do meaningful Broader Impacts, scientists new to evaluation requirements push back on these costs, but it is the price of doing business. Likewise, scientists who merely want to have program evaluation done to comply with NSF are missing the point. They will be less likely to benefit from the objective feedback that a solid evaluation plan has to offer.

Planning for evaluation should be an integral part of the proposal development process. Last-minute requests for evaluation services are not recommended; these sometimes result in a far higher estimate for services by the evaluator in order to cover contingencies or unexpected surprises from the project team. A good evaluator will invest some time in the project during proposal development with the hope that it will be funded. They typically provide a section in the proposal describing their evaluation plan and a budget and corresponding justification of the proposed expenditures. Program officers and reviewers at NSF typically will not push back on a well-developed evaluation plan that is up to 10 percent of total direct costs. Likewise, once funded, most (but not all) institutions accept the evaluator's proposed services without requiring a formal competitive bid process, although sometimes a sole-source justification may be required.

A good time to get the evaluator involved is when the research project team has reached consensus on the available resources, goals, objectives, activities, and desired outcomes and impacts. It may also be helpful to develop a graphical flow chart of these components, which has been termed a logic model (Diamond et al., 2009; Kellogg Foundation, 2006). The evaluator may also ask the project team what success will look like at the end of the project. In so doing, they can reverse-engineer the evaluation plan. It is also true that during the initial proposal development, many scientists have not fully thought through all of the potential impacts and outcomes. The evaluator can help facilitate this process as well.

Institutional Review Board

As part of the proposal development and evaluation process, the project team will need to have their data-collection protocols reviewed by an Institutional Review Board (IRB). Approval is required because, in order to conduct evaluation, data are typically gathered from human subjects. The IRB process was also created to protect the rights of human subjects, including confidentiality of personal information collected, analysis of risk to the subject (this is particularly the case in clinical trials, but not usually in scientific research evaluation), and informed consent. If minors are involved in the study, another layer of compliance is required – parental consent in addition to the child's own consent. Some K–12 school districts also have their own review process in addition to the IRB. These approval processes also have strict policies and procedures that restrict access to children's records without parental consent. Regardless of the target audience, IRB and other approvals can take time, typically a month or more for each review. Therefore, advanced planning is required in order to provide sufficient time for approvals before the study can begin.

For most of the kinds of evaluation done in support of scientific (i.e., non-medical) research, the risk is typically deemed minimal (e.g., UF IRB, 2017). Sometimes at the outset, once the evaluation protocol is described, subsequent IRB approval can be officially waived (i.e., in writing) unless significant changes occur during the course of the research. Like it or not, the sponsored research division at the host institution will require that the IRB process has been undertaken. The National Science Foundation will not fund a recommended proposal

Project evaluation

Fig. 18.1 The process of project evaluation, primarily following the model in informal STEM (e.g., Screven, 1990; Diamond et al., 2009).

until institutional IRB approval is granted (NSF, 2018h). Likewise, the evaluator will need to be a registered participant on the IRB-approved project. Because of the minimal risk involved in many evaluation studies that are not intended for research or dissemination, some changes are on the horizon, collectively termed "Common Rule," that may make this process easier in the future (Nichols et al., 2017) for the kinds of studies that would be proposed in most Broader Impacts plans.

The Evaluation Process

As an NSF research project passes through stages during its life cycle, so does the process of evaluation. Although there are some differences in the terminology and emphases (e.g., in K–12 versus informal projects), Screven (1990; also see Diamond et al., 2009) divided this cycle into front-end, formative, and summative evaluation (Fig. 18.1).

Front-End Evaluation

Prior to proposal submission, many projects will undergo a feasibility study, here called front-end evaluation (Diamond et al., 2009; IMLS, 2008; Screven, 1990). The purpose of this initial phase is to understand project, goals, and potential results (outcomes and impacts) – for example, the front-end evaluation of a shark exhibition at the Florida Museum of Natural History (Fig. 18.2). Another important consideration is to understand the target audience. Sometimes front-end evaluation can include a needs assessment (IMLS, 2008), meaning understanding prior research that informs the current project and how the new project advances the field. The front-end evaluation might also include a thorough literature review and some pilot testing or the production of initial data. In the project management cycle, this process relates to the planning and design phase (Fig. 17.1).

For the FOSSIL (2017) project, we conducted a front-end evaluation the year before we submitted the successful proposal. We analyzed the "ecosystem" of amateur paleontologists and fossil clubs in the United States and sent out an online survey to determine interest from our target audience. Because little was known previously about the status of fossil clubs and their members' interests in the United

Fig. 18.2 Front-end evaluation of a fossil shark exhibition during the feasibility and planning stages at the FLMNH. With funding from an Informal Science Education (ISE) supplement, this project ultimately helped to develop the *Megalodon* (2018) exhibition (FLMNH, Jeff Gage photo).

States, we published a report of the findings from our front-end evaluation (MacFadden et al., 2016). In another NSF program in graduate education, we submitted a request for funds for a front-end needs assessment to a program intended for small feasibility studies. We were interested in developing a program in which PhD students could train to develop hybrid careers in both K–12 and museums. That proposal did not review favorably for a variety of reasons. One reviewer commented that we had not yet documented the need, so it was a fatal flaw and too risky to be funded. This seemed counterintuitive because the program was intended to fund feasibility studies, but reviewers carry lots of weight when funding recommendations are made by the NSF program officer.

Formative Evaluation

Formative evaluation is done during project development and implementation. It seeks to make changes that will result in the improvement of the project or intervention. In the project management cycle (Fig. 17.1), monitoring and controlling can be a form of formative evaluation. This is the most common form of project evaluation (Lloyd et al., 2012) and includes both informal and formal K–12 learning settings. It is a feedback loop that is sometimes called iterative evaluation. No project is perfect from the start, and new ideas emerge during implementation. Thus, formative evaluation seeks to make improvements that will make a difference in terms of the overall success of the project.

Most scientists have done formative evaluation without knowing it. We devise an experiment, it does not work, then we make changes and do it over again, hopefully until it works better than before. In Fig. 18.3 we tested the *Megalodon* exhibition with museum visitors to understand if the text panels were effective in communicating the message and content. Oftentimes formative evaluation can have unanticipated outcomes: With the Megalodon exhibit panels we determined that certain combinations of red and green colors used

Fig. 18.3 Formative evaluation of the *Megalodon* panels. In addition to determining whether we were communicating the intended message to the visitor, we also found in previous versions that some of the color schemes (red–green) of the graphic panel mock-ups were not fully accessible to those with this kind of color-blindness.

by the graphic artists in the initial mock-ups could not be read by those who are blind in this color spectrum (which can affect up to almost 10 percent of males from northern European ancestry; National Eye Institute, 2018). If we had not done a formative evaluation of the prototypes and printed the panels, once on display they would have been less accessible to all of our visitors. Another common problem with exhibits is the font size on graphic panels. If fonts are too small, they will be difficult or impossible to read, particularly with elderly visitors with diminished vision. Books have been written on best practices for the use of fonts in museum exhibits (Serrell, 2015).

For this formative evaluation we conducted short interviews via a structured list of questions with a small segment of our museum visitors. Input during formative evaluation can come in many forms. It is best to concentrate on the intended audience for their feedback. In addition to interviews, data can be drawn from a variety of sources, including, for example, observation (without disturbing the visitor), survey questionnaires, and focus groups (Diamond et al., 2009).

In addition to receiving feedback from the target audience, another common type of formative evaluation is via focused input from experts or external advisors. During the Panama PIRE (2017) project we benefitted from our external advisory board, consisting of colleagues who were mostly familiar with the research goals of the project. At the end of Year 1, we were not finding the quantity of fossils along the Panama Canal that we had hoped. Our board advised us that rather than short "parachute" visits to the field (1–2 weeks), the project would greatly benefit from having a year-round crew monitoring new sedimentary exposures and finding many more important

Fig. 18.4 Left: Collecting fossils along the Panama Canal in 2012. This team was the first cohort of "boots-on-the-ground" field interns who lived in Panama while doing intensive field work. Right: Early Miocene (~19 million years old; MacFadden et al., 2014) fossil discoveries resulting from this cohort's field activities included this block of sediment containing a scapula of a rhinoceros (left, below scale bar) and jaw of an extinct bear-dog (right) (Stephanie Lukowski photos).

fossils. We responded by renting a house in Panama City and starting a residential internship program. Our person-days in the field increased dramatically, and over the next five years of the project we logged more than 8,000 person-days in the field, which was a significant increase from our baseline of a few hundred person-days logged in Year 1. This increased time in the field also yielded additional fossil discoveries that boosted the diversity of taxa represented in our collections (Fig. 18.4). Although it became clear that having an in-country, "boots-on-the-ground" presence in Panama would significantly improve productivity, and ultimately the success of the Panama PIRE, the spark of this idea did not come until our meeting with the external advisors at the end of Year 1.

Summative Evaluation

The third phase of this process is summative evaluation (Diamond et al., 2009; Fenichel & Schweingruber, 2010). This is typically conducted at the end of the project to understand whether the original project goals were successful. Summative evaluation can be done with mixed methods of data collection and analysis, including surveys, interviews, focus groups, and analysis of artifacts and deliverables (e.g., products such as lesson plans or exhibits). Sometimes summative evaluation uncovers unexpected outcomes, such as a part of an exhibit that is less effective than expected, which can then be changed (or "remedial" evaluation; *sensu* Screven, 1990) based on the new data. As already mentioned, some NSF programs, like Advancing Informal STEM Learning (NSF,

2017x), require that summative evaluation be conducted by an external evaluator to prevent bias or conflict of interest.

During the GABI RET (2017) project collecting fossils along the Panama Canal and localities in North America, we conducted evaluations at the end of each annual cohort. While, in a sense, they were summative for that year's cohort, our intention was to use the data collected to inform subsequent cohorts in a formative (iterative) way. After five cohorts spanning 2012–2016 we wanted to better understand the longer-term outcomes and impacts of this project. Very little had been published in the peer-reviewed literature about the impacts of Research Experiences for Teachers (RET) projects, including authentic, international research experiences. We structured the evaluation so that it would use validated instruments. In so doing, the data collected would be considered research grade – that is, the quality expected in peer-reviewed journals. An external evaluator was hired to conduct a summative evaluation that also had a longitudinal component, the latter investigating longer-term impacts on the earlier cohorts during the years since they actively participated in our project. The primary means of evaluation included an e-survey sent to two populations, the teachers and the scientists. These data were compiled and then the majority of the participants from the GABI RET cohorts 1–5 attended a summit (Fig. 11.5). During the summit we reviewed the survey results and the evaluators held focus group discussions with the participants. The final summative report was then completed within the next few months. This summative evaluation (Davey, 2017) documented many positive and long-term outcomes of this project, including significant unexpected benefit accrued back to the scientists. In addition, because we used validated instruments (surveys), we will be more likely to receive a more supportive response from an editor if we submit our findings to a science education journal.

Were You Successful?

Most scientists want to demonstrate that they were successful in their research. While some are satisfied if they have made discoveries, supported graduate students, presented talks, and published papers, this list is not sufficient in the current climate of Broader Impacts. Unless initial goals, objectives, and metrics are established, documenting the success of these activities can be elusive.

SMART and Meta-goals

Some scientists are content with stating ambiguous specific objectives and products. The concept of SMART "goals" (actually objectives; Doran, 1981) is relevant here. SMART goals (Inset 18.2) were developed as a tool in the business and nonprofit sectors and represent a best practice of management. They have more recently come

> **Path to success: Some important terms**
>
> **Goals:** Broad primary or strategic outcomes expected from project
> **Objectives:** Measurable or discrete steps of progress toward goal
> **Metrics:** Something that can be measured; performance indicators
> **Outputs:** Deliverables and activities that result from project
> **Outcomes:** Description of what happened, or project achievements
> **Impacts:** Did the project make a difference, or effect change?

Inset 18.1

into the academic world and are an effective way of planning project evaluation and documenting success.

Instead of "I will …

… work with teachers to develop lesson plans."

… work at underserved local schools."

… present results at a national conference."

> **SMART goals**
>
> Specific
> Measurable
> Achievable
> Relevant
> Time-bound

Inset 18.2

Use the following SMART framework – "I will …

… work with two middle-school teachers to develop four lesson plans, aligned with NGSS.

… work with two Title I schools with predominantly (> 80 percent) underserved learners.

… present project outcomes (poster or talk) at one science- and one education-focused professional meeting during the last year of the project funding period.

One would never know whether they had achieved the objectives stated in the first set of statements, whereas in the SMART framework the corresponding products and goals are accountable and evaluated as part of project outputs.

In our management plan for the Panama PIRE (2017; also see Chapter 17) project, we developed a long list of what were thought were SMART goals, meaning that at the outset they seemed to be attainable. During the course of any project, however, some goals and objectives may become unattainable. When I was a project officer at NSF, we had a goal that three-quarters of our proposals would be decided upon within six months of receipt. This provided us a relief valve for those that were problematic (e.g., took too long during negotiation). After returning from NSF and while writing our Panama PIRE Project Management Plan (2010) we developed a "meta-goal" – meaning an overall goal of the goals – that 80 percent of all of our individual goals stated in the written plan would be achieved. We were generally successful in this regard.

Another way of looking at this matter is that, within the framework of project management, the project should have flexibility and be nimble. Thus the project management plan is best considered a working document that responds to new opportunities, while also deleting initial goals and objectives that may no longer be feasible or important as the research develops. From a strategic point of view, sticking too rigidly to an initial plan and expecting to achieve 100 percent of a project's initial goals is a recipe for failure. This strategy adds unnecessary stress and expectations to the project team, but it does not have to unfold like this during the actual research.

Metrics and Outputs

Metrics are objectives, activities, deliverables, and outputs that can be measured in order to demonstrate efficacy or the overall success of a project. They are typically tied to project goals and the mission statement. These constitute the data that form a quantitative picture of a project. This process can include, for example, counts (e.g., of participants, publications, presentations), responses on surveys (e.g., Likert scales), and coding of variables. With regard to learning assessments, pre- and post-test methods can document knowledge gains. Quantitative data are typically tested for statistical significance and reported in tables or graphs. Within a mixed-methods evaluation design, the quantitative part includes the collection of metrical (statistical and measurable) data, whereas the qualitative side oftentimes includes observations and direct participant responses on open-ended surveys, interviews, or focus groups (e.g., Diamond et al., 2009). Both of these methods have value, and a mixed-methods approach is a dominant design for evaluation tools and protocols.

Outcomes and Impacts

For most projects we want to know "what happened?" both in terms of the research produced and Broader Impacts that were implemented. Likewise, did the project achieve the desired goals and objectives? An analysis of outcomes enhances understanding of the success of a project. Some outcomes are expected, while others that were unanticipated at the beginning of the project may be equally powerful contributions to the project. Outcomes can be evident at certain times during or after a project, and in some instances, such as disseminated research via peer-reviewed publications, may occur years after the funding period is over. Outcomes can also occur at different scales (Fig. 18.5): at the individual level, the group level, or in the case of citizen science, at the entire community level (Jordan et al., 2012). We predicted this spectrum of impact on different scales with our FOSSIL project (Fig. 18.5). When learning is involved, outcomes are typically tied to learning goals stated at the beginning of a project (Fenichel & Schweingruber, 2010). For example, if increased content knowledge about evolution is a learning goal of a project, then relevant learning gains can be assessed.

For more than a decade I co-taught an undergraduate course in evolution for science majors. More than once a student met with me to confess that although they learned the material for the test, and did well, they did not believe in evolution. This brings us to a discussion of impacts. If a goal of a project is to change acceptance and belief of a topic, this may be more complex than just an outcome, and also more challenging to evaluate. In the example about my student who successfully learned

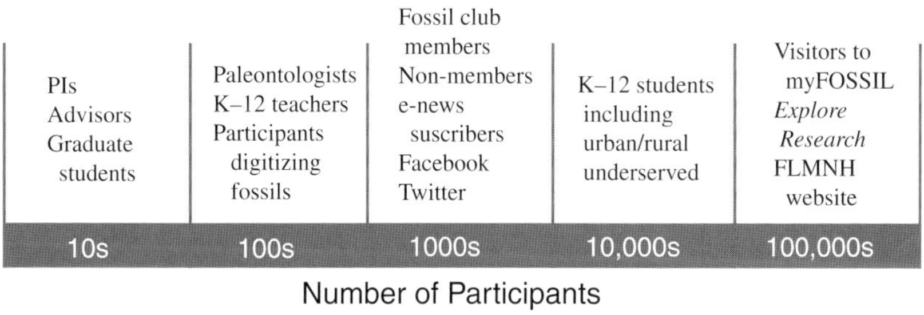

PIs Advisors Graduate students	Paleontologists K–12 teachers Participants digitizing fossils	Fossil club members Non-members e-news suscribers Facebook Twitter	K–12 students including urban/rural underserved	Visitors to myFOSSIL *Explore* *Research* FLMNH website
10s	**100s**	**1000s**	**10,000s**	**100,000s**

Number of Participants

Fig. 18.5 An a priori analysis for the FOSSIL (2017) project of potential impacts on different audiences and at different scales of participation (unpublished graphic modified from the original NSF proposal).

about evolution (outcome), this increased knowledge did not change their beliefs about acceptance of the theory (impact), whether or not the latter was a goal of the course. I do not push acceptance of evolution because it is better to present the topic and then let the students decide for themselves what they believe.

A common problem with controversial science topics (Leshner, 2010) in society is that although people may learn the content of these topics, they may bring misconceptions or strong opinions engrained in their prior knowledge that prevent acceptance of evolution or climate change, to give two current examples. With regard to these important topics that form scientific paradigms, strategic impact (e.g., Silverstein, 2008) is a way of "moving the needle" in a direction in which scientific evidence can sway public opinion, belief systems, and behaviors, if these are project goals.

Writing Reports and Publishing Your Results

A general expectation of most evaluations is that the results are prepared in a written report to be submitted to the researcher who commissioned the study. Although the data may be the intellectual property of the scientists, sometimes the instruments (i.e., evaluation tools) may be proprietary. The scientists will need to clarify this matter with the evaluator. After a preamble and introduction, a typical report includes the survey methods, data reporting and analysis, results, recommendations, and summary. These written evaluation reports are appended to the required annual (formative) and final (summative) reports submitted to NSF.

As described above, the purpose of evaluation and research are oftentimes different. Nevertheless, sometimes the results of a specific evaluation study may be generalizable to a larger audience. In such cases, the scientists and/or evaluators may decide to prepare an article to be submitted to a peer-reviewed journal. As also cautioned earlier, if the article is sent to a rigorous learning research journal, the project and evaluation design, as well as the instruments, may not meet the expectations of the reviewers and editors. If publication is, or may become, a goal of an evaluation study, to the extent possible, validated instruments should be used during data collection. Likewise, the choice of journal will be crucial. We were able to publish our front-end survey results about fossil clubs in a paleontological journal (MacFadden et al., 2016). However, this article would likely not have been appropriate for an education journal because the study goals and data collection were not focused on hypothesis-driven formal learning research methods. Likewise, in our case study of the *Fossils4Teachers!* professional development workshop for amateur paleontologists and teachers (Pirlo et al., 2018; Fig. 16.1), the journal allowed the evaluation to be aligned with the scope of the activity. In this case, the goal was to describe the activity and participant feedback rather than formally testing a learning research hypothesis.

Scientists who think that they can conduct learning research without expert advice will likely be, as I have been, disappointed by the outcome: papers rejected during peer review. In the event that a project expects some evaluation data of generalizable significance, it is best to involve a learning researcher or science educator. In so doing, eventual publication in peer-reviewed education journals will be a more viable option as well as add value to the project deliverables.

Concluding Thoughts

Similar to sustainability, project management, and Broader Impacts, evaluation is oftentimes a proposal element that many scientists would rather not address in a meaningful way. Many scientists are unfamiliar with how to proceed with this process. Organizations such as the Center for Advancement of Informal Science Education (CAISE, 2018) and the Visitor Studies Association (VSA, 2017) provide resources on evaluation. Several "user-friendly guides" (e.g., Frechtling et al., 2010) have been published to help novices in this field, and Diamond et al. (2009) have presented an accessible overview of evaluation in their book. These are just a few examples, of which many others exist in the evaluation world.

A key recommendation is that evaluation is best started early during the project and proposal development. The larger and more complex the project, the more important is the evaluation. Project teams that lack relevant expertise are well served to include a science education faculty member with experience in evaluation, or plan to seek the services of an external evaluator. If the evaluation process and plan are considered as a means of compliance (i.e., they are required by NSF), rather than truly understanding how the project can benefit from this process, then the PIs are missing a significant opportunity – one which they have to pay for regardless. Many times, formative evaluation will fine-tune the projects goals and objectives. Sometimes the evaluation will uncover unanticipated outcomes that may be of interest to a broader audience. In the latter case, the choice of an appropriate means of dissemination will be critical, including, for example, publishing in an appropriate peer-reviewed journal.

19 Wrap-Up, the Future, and Broader Impacts 3.0

Boilerplate and Templates: Worthless Time-Savers

I subscribe to a national listserv of the Broader Impacts community. As this book was nearing completion, I was fascinated to read a comment from a member of this listserv requesting a generic template statement that could be used as part of a Broadening Participation section in an NSF proposal. One member of the community responded as follows:

> My experience has been that the BP component is unique to each project as it is an integral part of that project. I have never used the same approach twice, because it would come off as tacked on. I have read many proposals when serving on panels in which the BP statements appear to have been stuck in to try and check off a requirement box. That rarely goes over well with the panel.
>
> (Lisa Kaczmarczyk, email communication, August 2018, reproduced with permission)

As you will understand from my introductory anecdote in Chapter 16 on Broader Impacts plans, I am not a fan of boilerplate and "one-size-fits-all" statements that are cut-and-pasted into NSF proposals. I am likewise supportive of Lisa's response to the request on the listserv. Simply put, the use of boilerplate text or templates should be avoided. Although they might save time, they compromise the integrity and originality of the plan, whether it be for Broader Impacts or Broadening Participation.

Introduction

A word cloud of about 30 words (Fig. 19.1) is a way to represent the content of this book. Along with the word cloud, some dominant themes emerge. These reappear in different contexts and chapters and thus provide the fabric that weaves together an understanding of Broader Impacts as envisioned here. Eight topics (Inset 19.1) are presented below because they rise to the top as the most important themes in the book.

Fig. 19.1 Word cloud generated from the text of this book for about 30 of the most frequent content words.

Dominant Themes

Innovation and Opportunity

Two decades ago, when the community of NSF proposers learned of the "new" Broader Impacts review criterion, many scrambled to figure out what they should say in their proposals. Professors said that they would infuse their research into the content of their courses. Others said that they would build a website about their research using the then-emerging internet. Although at the time these plans seemed like good strategies, they would not be viewed favorably today. The National Science Foundation (NSF) has evolved, now with the added expectation that to receive the highest ratings, both the research and Broader Impacts must be innovative. Therefore, it is best to propose something that has not already been done, but within a context that builds upon an existing body of knowledge. The thread of innovation should be woven into the fabric of the proposal.

Related to innovation is opportunity, which may require moving out of one's comfort zone. Many NSF proposers go back to the same programs year after year, only to be disappointed unless they have a compelling idea for a project. This fate is particularly apparent in some of the most competitive programs, which can have success rates of 10 percent or less. However, if a principal investigator (PI) is flexible

Dominant themes
• Innovation and opportunity
• Collaboration, networking, and sustainability
• Dissemination and communication
• Twenty-first-century technology
• Audience and demographics
• Gateway science and charismatic STEM
• Evaluation and strategic impact
• Social responsibility

Inset 19.1

enough to consider other options through which to accomplish their research, then many opportunities exist within other niches at NSF. For example, Research Experiences for Undergraduates (REUs), a cross-directorate program, promotes the advancement of knowledge and discovery in their panel competitions. In this niche, however, successful proposals will also focus on undergraduate mentoring and engagement during the research. In this case, broadening participation to recruit a diverse team will also be an important component of the project. In another NSF niche, if a PI is interested in developing an exhibit that focuses on the STEM content of their research, they can look for funds to develop these activities from programs in the education directorate. As is cautioned above, however, in this case it is advisable to collaborate with a science educator, one who knows how to frame learning research hypotheses and understands evaluation practices.

Over the next decade NSF will focus funding on the 10 Big Ideas (Chapter 7). These ideas encompass a broad swath within NSF's mission and priorities. If one's research relates to any of these 10 strategic initiatives, then the PI can align their projects within this framework of priorities. In so doing, proposals they submit may have a higher probability of funding within those NSF areas that anticipate future growth.

Collaboration, Networking, and Sustainability

The culture of science has changed over the past several decades. Single-investigator projects funded by NSF have given way, in many instances, to collaborative proposals involving multiple PIs and institutions, particularly for more complex research. This shift has evolved from factors such as increasing specialization, as well as the advent of multidisciplinary projects. Collaboration, therefore, is important to one's potential success as a scientist. There are challenges that we have discussed, including being sure to choose your collaborators carefully – that is, not jumping at any chance to be involved. Many long-time researchers have experienced collaborations gone sour. These can be stressful, are much less rewarding in terms of the social aspect of science, and can impact the overall quality and productivity of the research.

A related aspect of increasing collaboration is the need to network. In recent years NSF has supported this practice through funding research coordination networks (RCNs). Networking is fundamental to being a successful scientist in the twenty-first century. Early in my career, instead of going to all of the talks at professional meetings, I spent time chatting with colleagues. Without knowing it, I was networking and building professional connections that served me well later. Aspiring early-career scientists should know that these social interactions are, in many instances, as important as going to the talks at meetings. The networking will provide valuable support that potentially can enhance their chances of success during their careers.

The National Science Foundation is interested in making investments in research that can be sustained after their funding ends. In addition to the positive benefits for the research while it is being done, networking can also help the sustainability of projects after NSF funding ends. A sustainability plan is best developed and implemented during the course of the research. It can be enhanced by networking and collaborations. Research that waits to address sustainability until late in the project likely will not be as successful in the long run.

Dissemination and Communication

> Scientists need to learn to communicate better.
>
> Greenbaum and Gerstein (2018)

As described in Chapter 4, a distinction is made between dissemination and communication. Although the boundaries between these terms seem blurred to me, dissemination relates to transferring knowledge to one's community, such as through books, peer-reviewed publications, and presentations at professional meetings. The structure of these dissemination modes is engrained in the culture of what we do as scientists, including presenting our studies using the "scientific method," in much the same way as we were taught in school. Likewise, academic seminars are typically presented in the 50-minute hour, and talks at meetings in 15-minute slots. We tend to use jargon and an array of acronyms that limit the accessibility of our dissemination.

Communicating our science to broader audiences is another world, one that many scientists neither understand nor have been trained to do effectively. Yet, if we are to make an impact on society, effective communication is fundamentally important. Another related theme in communication is to know your audience. Talking to scientists is different from a general public lecture. Recall the inverted pyramid of communication (Fig. 4.1), in which the general public receives the big picture or take-home message first, with the talk then unfolding with evidence to back these up. In a certain sense, this also inverts the traditional scientific method.

Whether it be dissemination or communication, skills such as being able to encapsulate research in an elevator speech or clear graphics (e.g., a poster) will make the science content more accessible. Traditionally, these skills have not been taught within the halls of academe, although this culture is changing, particularly with the next generation of scientists. Effective communication of our research discoveries to the general public for the overall benefit of society is expected by NSF. This knowledge transfer is fundamental to the concept of Broader Impacts.

Twenty-First-Century Technology

> New technology is one of the most powerful drivers of scientific progress.
>
> Berg (2018b)

The National Science Foundation encourages funding of transformative discoveries. As has already been discussed, however, transformative research is difficult to recognize a priori. The business and for-profit sector teaches us that transformation is most easily assessed after the fact, with metrics and other indicators of success. Did the inventors of the internet fully realize its transformative impact and the extent to which it pervades modern society? It is clear now that the advent of cybertechnology has fundamentally transformed science and society in the twenty-first century. Examples of this transformation include communications, education, imaging, and data storage, all of which are the beneficiaries of this modern technology.

Cyberenabled communications have become commonplace in society. They allow us to connect with associates and build networks (e.g., communities of practice, or CoPs) that were previously impossible, particularly when participants live geographically separated from each other. We have also transformed K–12 education. Teachers and students now can have scientists visit their classrooms remotely (e.g., Skype a Scientist, 2018) or stream in real-time dynamic science content (Streaming Science, 2018), which oftentimes can then be digitally archived for future access. Likewise, the internet has enabled access to cloud resources, including massive databases that were previously impossible to access. There is no doubt that the "Big Data" revolution is here to stay and the STEM domain of data science is now a recognized field in academic circles. Harnessing this technology is one of the 10 Big Idea areas for future NSF investment.

In another technology transformation, digital imaging has revolutionized science. We now have access to digitized fossil specimens that can be imaged using sophisticated tools and devices, such as micro-CT scanners. The images produced from these are then post-processed using modern software. The images can be aggregated by large databases. The result of these innovations includes opening up museum research collections away from the realm of "dark data" (Marshall et al., 2018). These specimens can be 3D-printed for research, and also used for education in a way that was previously unimaginable. Rapid prototype devices, also called 3D-printers, were unknown in most educational institutions a decade ago. They are becoming more commonplace in schools and also have changed the way libraries attract visitors into maker spaces. As with all technologies, after an initial wave of innovation, 3D-printers will become commonplace. In the future they will likely become accepted as helpful appliances in many walks of life. Future innovations will come from how they are used, and less from the novelty of the 3D appliance itself. Companies that develop 3D-printers aspire for them to be simple to use, with little training or maintenance required. My experience, however, has been that they still break down and require a learning curve daunting to some; this will hopefully change in the future. In another world, VR (virtual reality) and AR (augmented reality) are within the innovation cycle in which experimentation is widespread, and transformative discoveries and applications are on the horizon. Scientists wanting to develop modern Broader Impacts are thus well served to consider how they can harness the power of technology.

Audience and Demographics

Scientists, particularly those who have careers in higher education, have a culture in which their audiences are relatively homogeneous, or so they think. They present their research to colleagues at seminars and professional meetings, most of whom speak the same language of science with specific jargon and acronyms. For their undergraduates, most academics realize that an initial barrier exists with technical terminology until the students are able to assimilate the new language of the specific scientific domain. However, the expectation is that students need to come up to the professor's level. The thought of "dumbing down" (a loathsome term) science content so that it is more easily understood by the novice learner is anathema. This also is a myopic point of view that needs to change.

In the real world, the general public is an entirely different place for science communication. Whether it be a talk to a civic organization, science café, museum tour, or K–12 classroom visit, to name just a few, these audiences can be fundamentally different from the ivory tower. If one cares about accessibility in science communication to public (non-scientific) audiences, then one should seek to understand them. Although it is impossible to fully predict, if we can anticipate the audience's prior knowledge and topics that will engage their interest, this planning will likely result in a more successful talk. Likewise, audiences are hardly monolithic in their knowledge, interests, and demographics. An effective speaker, one who really cares about presenting an engaging talk, will understand these factors and prepare accordingly. Despite advance preparation, however, this is not a failsafe method – for example, when I presented a talk at a fossil festival families with infants in strollers started heading for the exit as I began to explain the geochemistry of fossil horse teeth. A wise colleague at NSF taught me that one of the most important things a scientist needs to know is their intended audience. This is good advice for any scientist wanting to do effective outreach.

Gateway Science and Charismatic STEM

The days of rote memorization of facts to learn science are hopefully behind us. Science, and in a larger context STEM, should be communicated as vibrant and exciting topics of current relevance to students and society at large. Some topics, like dinosaurs, tend to be intrinsically more engaging and thus are a charismatic gateway to learning. Moreover, when multiple charismatic domains of STEM work together, such as when using 3D-printed Megalodon teeth (Chapter 3), it is understandable that these are even more exciting to learners with diverse interests.

Catalyzed by the initial "hook," charismatic STEM thus provides a gateway for learning a variety of topics. In my domain, fossils and paleontology might be viewed as a rather narrow discipline, one that is stuck in the dusty corridors of museums. On the contrary, paleontology has broad applicability to science domains including anthropology, biology, chemistry, earth and environmental sciences, and physics. In addition, paleontology and any of the related fields are excellent gateways to understanding the process (or nature) of science. These endeavors can also lead to a better understanding of potential careers, which are less well known to young learners of the next generation. Finally, through the lens of charismatic gateways, we also find opportunity to discuss fundamentally important topics in modern society, such as evolution and climate change.

So far we have spoken about the STE in STEM, but the M, or math, is oftentimes taught independently. Nevertheless, it should not be left out of this discussion. Many studies have developed programs or lessons in which math can be learned through science, such as on a museum trip to measure specimens in hands-on displays. Furthermore, STEAM (including art) can also be used because of the added benefit of engaging participants with an artistic flare and skill set.

Evaluation and Strategic Impact

Evaluation is frequently either misunderstood, or disparaged, by scientists. Some of these researchers are content to receive funds, comply with the Broader Impacts, and be done with this part of the project – with little regard for accountability. Nevertheless, would it not be better to know that we are successful? The landscape is changing because NSF increasingly expects accountability, and it is best to adapt, or become extinct.

For a long time, educators have embraced evaluation as part of their culture. In this regard they are way ahead of scientists. Educators constantly evaluate programs to determine if they have been effective. Without this evidence, it is impossible to know if funds have been well spent toward strategic impacts such as student achievement. In terms of scientific research, the bar continues to be raised in NSF-funded programs. Evaluating the efficacy of NSF's investments is here to stay, and will likely become more structured and sophisticated in the future. Decades ago, when I started to receive NSF funding, neither Broader Impacts nor evaluation were part of the deal. At the present time, larger projects funded in NSF's research directorates are rapidly evolving a culture of evaluation and accountability.

With the additional challenge of evaluation, many scientists do not know how to do it, or who will do it. Unless we come from the education world, we are simply not trained to do evaluation; this practice is a science unto itself. Hastily developed surveys (lacking a systematic research design and Institutional Review Board approval) may work as a basic strategy. These surveys, however, will almost never be acceptable in professional presentations and to generate results intended for peer-reviewed publications. Like Broader Impacts, a well-planned evaluation requires a budget, sometimes as much as 10 percent of the project's direct costs. This investment frequently comes as a surprise to the uninformed.

From the philosophical point of view of being a scientist, it is important to know if you have made a difference in the world, for the better – this is at the core of long-term outcomes, or strategic impacts. For example, did the Broader Impacts activities change attitudes about topics of modern relevance, such as evolution or climate change? The best way to document positive gains and strategic impact is via a rigorous evaluation protocol. As someone who was new to evaluation a decade ago, I have come to value this process as tangible evidence of the overall success of a project.

Social Responsibility

Returning to the introductory anecdote in Chapter 1, I ask my students why they want to be a scientist. A common response is that they want to make the world a better place and have a positive impact on society. Without explicitly discussing the underlying reason for this response, it is clear that these students are motivated through the lens of social responsibility. For the other students who are inwardly focused and want to discover something important, they likewise should understand social responsibility. This understanding is even more important for those of us who have the luxury of pursuing a rewarding career of basic, or pure, science supported by taxpayers' dollars. In the bigger picture, everything that we do as scientists, including when we are developing Broader Impacts, is inherently guided by the conceptual frameworks of social responsibility and overall benefit to society.

The Emerging Paradigm: Broader Impacts 3.0

In Chapter 2, the assertion is made that, although not called such, the concept of societal benefit ("welfare") that ultimately led to Broader Impacts has been part of NSF's culture since it was founded in 1950. This concept has evolved over the past 70 years, always in the direction of increased visibility and importance. To the chagrin of some scientists, in 1997 Broader Impacts were formalized as one of the two proposal review criteria, along with the Intellectual Merit of the proposed science. In a paper that uses the phrase "Broader Impacts 2.0," Frodeman et al. (2013) describe a transformation of the proposal review process based on: (1) the America COMPETES Act of 2010 passed by the US Congress; (2) the National Science Board's subsequent recommendations in 2011; and (3) NSF's updated Proposal & Award Policies & Procedures Guide published in October 2012 (NSF, 2012a). They argue that, for the first time, Broader Impacts is set on an equal footing with Intellectual Merit as the two proposal review criteria. Since that time, the culture has continued to evolve so that, in most proposal review panels, Broader Impacts are taken more seriously in funding recommendations. This suggests that Broader Impacts will inevitably become of increased importance in the future.

Broader Impacts 3.0: Broadening Participation

> Guided by the Strategic Plan, NSF established a performance area focused on broadening participation: to expand efforts to increase participation from underrepresented groups and diverse institutions throughout the United States in all NSF activities and programs.
>
> NSF (2018i)

Vannevar Bush (1945) emphasized the importance of diversity in science and technology, not just for the generation of ideas, but also for the development of the workforce in the United States. This likewise has been a core value of NSF since its inception. Many NSF programs have supported diversity, inclusion, and equity in one way or another over the years, and this has evolved into the more recent concept of Broadening Participation. Of direct relevance here, during the past decade the INCLUDES program (NSF, 2017n) is increasing NSF's focus on: (1) diversity, equity, and inclusion in funded research projects; and (2) strategic outcomes for STEM workforce development. It is a centerpiece of NSF's Broadening Participation portfolio (NSF, 2018j) that includes about two dozen individual programs and initiatives, some of which have been around for a while, whereas others like INCLUDES have been implemented more recently. This Broadening Participation portfolio will fundamentally transform who participates in science and STEM and will become part of the core culture of federally funded research during the twenty-first century. As with Broader Impacts in recent decades, scientists who want to write a more competitive NSF proposal will now need to think about how their research can integrate diversity, equity, and inclusion in an impactful way. While Broadening Participation has not yet achieved parity with Intellectual Merit and Broader Impacts, it likely will play an increasing role within the framework of NSF-supported research.

Final Thoughts

> An enlightened citizenry is indispensable for the proper functioning of a republic.
>
> Thomas Jefferson (in Reddy, 2009)

The overarching goal of this book is to present NSF's Broader Impacts, and in a larger sense how these and similar activities can benefit society. From a pragmatic point of view, this book is also intended for scientists wanting to be more successful in preparing NSF proposals. In order to provide context for these goals, the development of NSF, science, and now STEM since the middle of the twentieth century is provided as a conceptual and historical framework. To this are added specific examples and case studies that illuminate representative Broader Impacts activities. It is my hope that scientists previously unfamiliar with, or uninterested in, this topic will find this book useful and perhaps even inspire innovation as part of their own professional activities.

The National Science Foundation expects to receive more than 50,000 proposals in fiscal year 2019 (NSF, 2018b), of which no more than one-quarter will likely be funded. These proposals will directly impact more than 350,000 people, including senior and early-career professionals, postdocs and students, K–12 teachers and students, and lifelong learners in the United States. As envisioned by Vannevar Bush (1945), NSF has been the fundamental driver of basic science and technology (now called STEM) for the past 70 years and will continue in this role in the foreseeable future.

In terms of strategic priorities, scientists would be wise to determine whether their research aligns with one or more of the 10 Big Ideas for NSF's investments for the foreseeable future (Inset 7.5). Likewise, in addition to the current emphasis on both Intellectual Merit and Broader Impacts, Broadening Participation (INCLUDES is one of the 10 Big Ideas) has also become of greater importance in the development of research plans and proposals. Three-quarters of a century has shown us that Broader Impacts are here to stay. They have been, and will continue to be, part of NSF's mission, and as such a core value of basic research supported by the US government. Broader Impacts are fundamentally integral to the best science being funded these days. Scientists in the twenty-first century will engage in innovative research activities and society will ultimately benefit from NSF's investment in Broader Impacts.

References

[T] = URL truncated, but the reference can be reached directly by typing in the reference title, author, and/or date in the website search function.

AAAS (American Association for the Advancement of Science). 2010. Communicating science tools for scientists and engineers. Workshop presented at the AAAS Annual Meeting, San Diego, California, 18 February 2010.

AAAS (American Association for the Advancement of Science). 2017. Taking a poke at peer review. *Science*, 358:429.

AAAS (American Association for the Advancement of Science). 2018a. French balk at journal fees. *Science*, 360:11.

AAAS (American Association for the Advancement of Science). 2018b. Graduate STEM fellows in K–12 education. Accessed 9 April 2018 at: www.gk12.org.

AAAS (American Association for the Advancement of Science). 2018c. Trust in scientists remains high. *Science*, 359:851.

AAM (American Association of Museums). 2010. Demographic transformation and the future of museums. Accessed 31 December 2018 at: www.aam-us.org/wp-content /uploads/2017/12/Demographic-Change-and-the-Future-of-Museums.pdf.

AAM (American Alliance of Museums). 2016. Volunteers and museum labor. Accessed 14 June 2018 at: www.aam-us.org [T].

AAMV (American Association of Museum Volunteers). 2012. Standards and best practices for museum volunteer programs. Accessed 14 June 2018 at: https:// aamv.wildapricot.org [T].

AAUW (American Association of University Women). 2015. Even in high-paying STEM jobs, women are being shortchanged. Accessed 26 November 2018 at: www.aauw.org [T].

Abramson, L. 2007. Sputnik left legacy for U.S. science education. Accessed 13 June 2017 at: www.npr.org [T].

Adam, T. 2019. The future of MOOCs must be decolonized. Accessed 27 May 2019 at: www.edsurge.com.

Agrawal, A. A. 2005. Corruption of journal impact factors. *Trends in Ecology and Evolution*, 20:157.

Aiken, J. W. 2006. What's the value of conferences? *The Scientist*. Accessed 7 November 2017 at: www.the-scientist.com/ [T].

Alachua Astronomy Club. 2018. Solar Walk. Accessed 8 June 2018 at: www .alachuaastronomyclub.org/ [T].

Alberts, B. 2010. Promoting scientific standards. *Science*, 327:12.

Alberts, B. 2013a. Impact factor distortions. *Science*, 340:787.

Alberts, B. 2013b. Prioritizing science education. *Science*, 340:249.

Alizon, S. 2018. Inexpensive research in the open-access era. *Trends in Ecology & Evolution*, DOI:https://doi.org/10.1016/j.tree.2018.02.005.

Allen, S. 2004. Designs for learning: studying science museum exhibits that do more than entertain. *Science Education*, 88 Supplement 1:S17–S33.

Alphamu57. 2018. Project management flow chart. Accessed 26 July 2018 at: https://commons.wikimedia.org/w/index.php?curid = 3976594.

Altmetric. 2017. Accessed 20 September 2017 at: www.altmetrics.com.

Amazon. 2018. [Review of] *Fossil horses: Systematics, paleobiology, and evolution of the Family Equidae.* Accessed 2 September 2018 at: www.amazon.com [T].

America Competes Act. 2010. America Competes Reauthorization Act of 2010. Accessed 31 December 2018 at: www.congress.gov/111/plaws/publ358/PLAW-111publ358.pdf.

American Astronomical Society of the Pacific. 2017. Accessed 7 November 2017 at: www.astrosociety.org.

AMNH (American Museum of Natural History). 2017. Travelling exhibitions: Darwin. Accessed 12 December 2017 at: www.amnh.org [T].

Aslam, S. 2017. Twitter by the numbers: stats, demographics & fun facts. Accessed 16 September 2017 at: www.omnicoreagency.com/twitter-statistics.

ASTC (Association of Science – Technology Centers). 2018. Roy L. Schafer Leading Edge Awards. Accessed 15 October 2018 at: www.astc.org [T].

AstroDance. 2018. About AstroDance. Accessed 28 August 2018 at: http://astrodance.rit.edu/node/29.

Audubon. 2018. Equity, diversity and inclusion at Audubon. Accessed 15 September 2018 at: www.audubon.org [T].

Ball, P. 2005. Index aims for fair ranking of scientists. *Nature*, 436:900.

Barnosky, A. D., P. L. Koch, R. S. Feranec, S. L. Wing, & A. B. Shabel. 2004. Assessing the causes of Late Pleistocene extinctions on the continents. *Science*, 306:70–75.

Baskas, H. 2016. Tuesday is Dino-day at Chicago's O'Hare airport. *USA Today*. Accessed 9 June 2018 at: www.usatoday.com [T].

Bell, P., B. Lewenstein, A. W. Shouse & M. A. Feder (eds.). 2009. *Learning Science in Informal Environments*: *People, Places, and Pursuits*. Washington, DC: National Academies Press.

Benton, T. H. 2010. Getting real at natural-history museums. *Chronicle of Higher Education*. Accessed 9 June 2018 at: www.chronicle.com [T].

Berg, J. 2017a. Science, big and small. *Science*, 358:1504.

Berg, J. 2017b. Preprint ecosystems. *Science*, 357:1331.

Berg, J. 2018a. Transparent author credit. *Science*, 359:961.

Berg, J. 2018b. Revolutionary technologies. *Science*, 361:827.

Bidwell, A. 2015. STEM workforce no more diverse than 15 years ago. *U.S. News & World Report*. Accessed 21 December 2017 at: www.usnews.com [T].

Bik, H. M. & M. C. Goldstein. 2013. An introduction to social media for scientists. *PLoS ONE*, 11:e1001535.

Biomuseo. 2018. Accessed 22 July 2018 at: www.biomuseopanama.org/en.

Blakey, B. 2015. Nonformal education. Accessed 9 June 2018 at: http://etec.ctlt.ubc.ca/510wiki/Nonformal_Education.

Blank, R., R. J. Daniels, G. Gilliland, et al. 2017. New data inform career choices in biomedicine. *Science*, 358:1388–1389.

Blickenstaff, J. C. 2005. Women and science careers: leaky pipeline or gender filter? *Gender and Education*, 17:369–386.

Bloom, B. S., M. D. Engelhart, E. J. Furst, W. H. Hill, & D. R. Krathwohl. 1956. *Taxonomy of Educational Objectives: The Classification of Educational Goals. Handbook I: Cognitive Domain*. New York: David McKay Company.

Boch, F. & A. Piolat. 2005. Note taking and learning: a summary of research. Accessed 17 October 2018 at: https://wac.colostate.edu [T].

Bohannon, J. 2016. Hate journal impact factors? New study gives you one more reason. *Science*, DOI:10.1126/science.aag0643.

Bokor, J., J. Broo & J. Mahoney. 2016. Using fossil teeth to study the evolution of horses in response to climate change. *American Biology Teacher*, 78:166–169. DOI:10.1525/abt.2016.78.2.166.

Bollyky, T. J. & P. L. Bollyky. 2012. Obama and the promotion of international science. *Science*, 338:610–612.

Bolton, A., C. Popson, & G. Roberson. 2018. Engaging family groups in learning about evolution using 3D printed fossils. *Geological Society of America Abstracts with Programs*. 50(6), DOI:10.1130/abs/2018AM-321306.

Bonney, R., H. Ballard, R. Jordan, et al. 2009a. *Public Participation in Scientific Research: Defining the Field and Assessing Its Potential for Informal Science Education*. Washington, DC: CAISE.

Bonney, R., C. B. Cooper, J. Dickinson, et al. 2009b. Citizen science: a developing tool for expanding science knowledge and scientific literacy. *Bioscience*, 59:977–984.

BOP (Billion Oyster Project). 2018. Accessed 23 June 2018 at: https://billionoys terproject.org.

Bornmann, L. 2014. Do altmetrics point to a broader impact of research? An overview of benefits and disadvantages of altmetrics. *Journal of Informetrics*, 8:895–903.

Borysiewicz, L. & N. Buckley. 2014. Festival lessons. *Science*, 343:949.

Box Office Mojo. 2017. *The Martian*. Accessed 27 December 2017 at: www .boxofficemojo.com [T].

Braaten, D. 2017. Let experts judge research potential. *Science*, 358:731.

Brainard, J. 2008. Tell us the "broader impacts" of your science. *Chronicle of Higher Education*. Accessed 9 June 2017 at: www.chronicle.com [T].

Brainscoop. 2018. Accessed 17 September 2018 at: www.youtube.com /thebrainscoop.

Bransford, J. D., A. L. Brown, & J. W. Pellegrino (eds.) 2000. *How People Learn: Brain, Mind, Experience, and School*. Washington, DC: National Academy Press. Accessed 4 September 2017 at: www.nap.edu/catalog/9853.

Britt, R. R. 2006. Beyond the geeks: 60 million Americans labelled "intellectually curious." *Live Science*. Accessed 26 December 2017 at: www .livescience.com [T].

Brody, R. 2015. What's missing from "The Martian." *The New Yorker*, Accessed 6 December 2017 at: www.newyorker.com/culture/ [T].

Brooks, M. 2012. Why the scientist stereotype is bad for everybody, especially kids. *Wire*. Accessed 21 December 2017 at: www.wired.com/2012/06/opinion-scientist-stereotype.

Brown, J. S. & P. Duguid. 2000. Balancing act: how to capture knowledge without killing it. *Harvard Business Review*. Accessed 15 September 2018 at: https://hbr .org [T].

Budiansky, S. 1996. Fossils? They can dig it. *U.S. News & World Report*, 120 (24):70–72.

Budiansky, S. 2013. *Blackett's War: The Men Who Defeated the Nazi U-Boats and Brought Science to the Art of Warfare*. New York: Vintage Books.

Bush, V. A. G. 1945. *Science: The Endless Frontier*. Washington, DC: United States Government Printing Office. Accessed 7 May 2017 at: www.nsf.gov/about/his tory/nsf50/vbush1945.jsp.

CAISE. 2018. Center for Advancement of Informal Science Education. Accessed 9 June 2018 at: www.informalscience.org.

Cambridge, H. 2016. Why do people take part in citizen science? Stockholm Environmental Institute. Accessed 23 June 2018 at: www.sei.org [T].

Cambridge Core. 2018. Paleobiology. Accessed 31 December 2018 at: www.cambridge.org/core/journals/paleobiology.

Cambridge Science Festival. 2018. Accessed 31 December 2018 at: www.sciencefestival.cam.ac.uk.

Cao, C. & R. P. Suttmeier. 2017. Challenges of S&T system reform in China. *Science*, 335:1019–1021.

Cardillo, J. 2015. The Americans with Disabilities Act: 25 years later. BrandeisNOW. Accessed 20 October 2018 at: www.brandeis.edu [T].

Carson, S. 2011. MIT OpenCourseWare introduces courses designed for independent learners. *MIT News*. Accessed 31 December 2018 at: http://news .mit.edu.

Cartlidge, E. 2018. The light fantastic. *Science*, 359:382–385.

Cassuto, L. 2016. *The Graduate School Mess*. Cambridge, MA: Harvard University Press.

CCMR (Cornell Center for Materials Research). 2017. Scientific poster design. Accessed 2 September 2017 at: http://hsp.berkeley.edu/sites/default/files/Sci entificPosters.pdf.

Chick, N. 2017. Learning styles. Vanderbilt University Center for Teaching. Accessed 4 September at: https://cft.vanderbilt.edu/guides-sub-pages [T].

Chicone, S. J. & R. A. Kisel. 2014. *Dinosaurs and Dioramas: Creating Natural History Exhibitions*. New York: Routledge.

Citizen Science. 2018. Toolkit: how to step by step. Accessed 23 June 2018 at: www .citizenscience.gov/toolkit/howto.

Clifford, P. S., C. N. Fuhrmann, B. Lindstaedt, & J. A. Hobin. 2014. Getting the mentoring you need. *Science*, DOI:10.1126/science.caredit.a1400027

Coburn, T. 2014. Wastebook: what Washington doesn't want you to read. Accessed 1 June 2017 at: http://coburn.library.okstate.edu [T].

Cohn, J. 2008. Citizen science: can volunteers do real research? *Bioscience*, 58:192–197.

Common Core. 2018. Common Core State Standards Initiatives. Accessed 5 April 2018 at: www.corestandards.org.

Conant, J. 2002. *Tuxedo Park: A Wall Street Tycoon and the Secret Palace of Science that Changed the Course of World War II*. New York: Simon & Schuster.

Connected Science Learning. 2018. Accessed 9 June 2018 at: http://csl.nsta.org.

Conner, T., D. Capps, B. Crawford, & R. Ross. 2013. Fossil finders: engaging all of your students using project-based learning. *Science Scope*, 36:69–73.

Cook, J. 2018. Importance of teaching science in elementary school. Accessed 5 April 2018 at: https://classroom.synonym.com [T].

Coons, C. 2017. Scientists can't be silent. *Science*, 357:431.

Cornell Laboratory of Ornithology. 2018. Accessed 23 June 2018 at: www.birds.cornell.edu.

Cornell University Library. 2018. Measuring your research impact: i10-Index. Accessed 29 August 2018 at: http://guides.library.cornell.edu [T].

Costello, M. J., W. K. Michener, M. Gahegan, Z.-Q. Zhang, & P. E. Bourne. 2013. Biodiversity data should be published, cited, and peer reviewed. *Trends in Ecology & Evolution*, 28:454–461.

Costello, T. 2018. NextGen postdocs. *Science*, 360:689.

Coursera. 2018a. Accessed 27 April 2018 at: www.coursera.org.

Coursera. 2018b. Dino 101 [MOOC]. Accessed 25 April 2018 at: www.coursera.org/learn/dino101.

Cousins, C. 2015. 10 tips for perfect poster design. Accessed 2 September at: https://designshack.net [T].

Cox, D. T. C., H. L. Hudson, D. F. Shanahan, R. A. Fuller, & K. J. Gaston. 2017. The rarity of direct experiences of nature in an urban population. *Landscape and Urban Planning*, 160:79–84.

Cox, R., K. Hunter, K. Cook, & S. B. Bush. 2019. Problem-based paleontology: a STEAM exploration for fourth graders. *Science & Children*, 56(5):42–48.

CPALMS. 2018. Accessed 11 November 2018 at: www.cpalms.org/Public.

CPET (Center for Precollegiate Education and Training), University of Florida. 2018. Accessed 5 April 2018 at: www.cpet.ufl.edu.

Crippen, K. J., S. Ellis, B. A. Dunckel, A. J. W. Hendy, & B. J. MacFadden. 2017. Seeking shared practice: a juxtaposition of the attributes and activities of organized fossil groups with those of professional paleontology. *Journal of Science Education and Technology*, DOI:10.1007/s10956-016-9627-3.

Cross, R. 2017. The inside story on 20,000 vertebrates. *Science*, 357:742–743.

Customer Magnetism. 2017. What is an infographic? Customer Magnetism digital and social marketing. Accessed 4 September 2017 at: www.customermagnetism.com [T].

Dahlstrom, M. F. 2014. Using narratives and storytelling to communicate science with nonexpert audiences. *Proceedings of the National Academy of Sciences*, 111, supplement 4:13614–13620.

Davey, B. 2017. Panama GABI RET 5-year summative evaluation. Unpublished report. Accessed 27 February 2018 at: www.gabiret.org.

Davidson, R. 2015. Even in high-paying fields, women are shortchanged. American Association of University Women. Accessed 26 December 2017 at: www.aauw.org [T].

Davis, G. 2005. Doctors without orders: highlights of the Sigma Xi postdoc survey. Accessed 3 September 2018 at: www.sigmaxi.org [T].

Derouin, S. 2018. Fieldwork among the pixels: virtual and augmented reality diversify geoscience education. *Earth*, May/June:72–79.

DeSantis, L. R. G. 2009a. Teaching evolution through inquiry-based lessons of uncontroversial science. *American Biology Teacher* 71 (2):106–111, DOI:10.1662/005.071.0211.

DeSantis, L. R. G. 2009b. Straight from the mouths of horses and tapirs: using fossil teeth to clarify how ancient environments have changed over time. *Science Scope* 32(5):18–24

DeSantis, L. R. G. 2017. I'm not your mother. *Science*, 359:690.

Detsky, A. S. & M. O. Baerlocher. 2007. Academic mentoring: how to give it and how to get it. *Journal of the American Medical Association*, 16:2134–2136.

De Winter, J. C., Zadpoor, A. A., & Dodou, D., 2014. The expansion of Google Scholar versus Web of Science: a longitudinal study. *Scientometrics*, 98 (2):1547–1565.

Diamond, J., J. J. Luke, & D. H. Uttal. 2009. *Practical Evaluation Guide*. Lanham, MD: Altimira.

Dickinson, J. & R Bonney (eds.). 2012. *Citizen Science: Public Participation in Environmental Research*. Ithaca, NY: Cornell University Press.

Dickinson Jr., J. C. (principal investigator) 1965. Florida State Museum application to the National Science Foundation: A grant to construct a specialized research facility for natural and social sciences. University of Florida, Gainesville, June 1965. FLMNH Archives, Office of the Director.

Dickinson Jr., J. C. 1966. Florida State Museum director's report. Unpublished, University of Florida, 1964–1966. FLMNH Archives, Office of the Director.

Dietl, G. P. & K. W. Flessa. 2011. Conservation paleobiology: putting the dead to work. *Trends in Ecology & Evolution*, 26:30–37, DOI:10.1016/j.tree.2010.09.010.

Dietl, G. P. & K. W. Flessa (eds.). 2018. *Conservation Paleobiology: Science and Practice*. Chicago, IL: University of Chicago Press.

Dijkgraaf, R. 2017. *The Usefulness of Useless Knowledge: Abraham Flexner with a Companion Essay by Robert Dijkgraaf*. Princeton, NJ: Princeton University Press.

Dolan, E. & K. Tanner. 2005. Moving from outreach to partnership: striving for articulation and reform across the K–20 + science education continuum. *Cell Biology Education*, 4:35–37.

Doran, G. T. 1981. There's a S.M.A.R.T. way to write management's goals and objectives. *Management Review: AMA FORUM*, 70 (11): 35–36.

Dorfman, E. 2017. *The Future of Natural History Museums*. London: Routledge.

Doubleday, Z. A. & S. D. Connell. 2017. Publishing with objective charisma: breaking down science's paradox. *Trends in Ecology & Evolution*, 32:803–805.

Dubois, M. 2014. STEM leaky pipeline. STEM Oregon. Accessed 12 May 2017 at: http://stemoregon.org [T].

Duncan, R. G. & A. E. Rivet. 2013. Science learning progressions. *Science*, 339:396–397.

Durant, J. 2013. The role of science festivals. *Proceedings of the National Academy of Sciences*, 110:2681

Dyehouse, J. 2011. "A textbook case revisited": visual rhetoric and series patterning in the American Museum of Natural History's horse evolution displays. *Technical Communication Quarterly*, 20:327–346.

eBird. 2018. Accessed 7 July 2018 at: https://ebird.org/home.

edX. 2018. Accessed 27 April 2018 at: https://learn.edx.org.

EEOC (Equal Employment Opportunity Act). 2017. The Americans with Disabilities Act of 1990. Accessed 26 December 2017 at: www.eeoc.gov [T].

Effland, A. B. W. & K. Kassel. No date. Hispanics in rural America. US Department of Agriculture. Accessed 22 December 2017 at: www.ers.usda.gov [T].

Ellwood, E. R., P. Kimberly, R. Guralnick, et al. 2018. Worldwide engagement for digitizing biocollections (WeDigBio): the biocollections community's citizen-science space on the calendar. *Bioscience*, 68:112–124.

Emery-Wetherell, M. M., B. K. McHorse, & E. B. Davis. 2017. Spatially explicit analysis sheds new light on the Pleistocene megafaunal extinction in North America. *Paleobiology*, DOI:10.1017/pab.2017.15.

English, L. D. 2016. STEM education K–12: perspectives on integration. *International Journal of STEM Education*, 3:3, DOI:10.1186/s40594-016-0036-1.

Enserink, M. 2018. European funders seek to end reign of paywalled journals. *Science*, 361:957.

Eshach, H. 2006. Bridging in-school and out-of-school learning: formal, non-formal, and informal education. *Journal of Science Education and Technology*, 16:171–189.

Evans, E. M. 2006. Intuition and understanding: how children develop evolution concepts. *ASTC (Association of Science – Technology Centers) Dimensions*, March/April:11–13.

Evans, E. M. 2013. Conceptual change and evolutionary biology: taking a developmental perspective. In S. Vosniadou (ed.), *International Handbook of Research on Conceptual Change*, 2nd edition (pp. 220–239) New York: Routledge.

Everett, K. 2011. *Designing the Networked Organization*. New York: Business Expert Press.

Exploratorium. 2018. Accessed 9 June 2018 at: www.exploratorium.edu/about/history.

Falk, J. H. 2001. Free-choice science learning: framing the discussion. In J. Falk, E. Donovan, & R. Woods (eds.), *Free-Choice Science Education: How We Learn Outside of School* (pp. 3–20). New York: Teachers College Press.

Falk, J. H. & L. D. Dierking. 2002. *Lessons without Limit: How Free-Choice Learning Is Transforming Education*. Walnut Creek, CA, Altamira.

Falk, J. H. & L. D. Dierking. 2010. The 95 percent solution. *American Scientist*, 98:486–493.

Farrell, B. & M. Medvedeva. 2010. Demographic transformation and the future of museums. American Association of Museums. Accessed 9 December 2017 at: www.aam-us.org/docs/center-for-the-future-of-museums/demotransaam2010.pdf.

FCAT (Florida Comprehensive Assessment Test). 2018. Florida Department of Education. Accessed 4 April 2018 at: www.fldoe.org/accountability/assessments.

Fenichel, M. & H. A. Schweingruber. 2010. *Surrounded by Science: Learning Science in Informal Environments*. Washington, DC: National Academies Press.

Ferguson, M., M. Minarchek, N. Porticella, & R. Bonney. 2014 *User's Guide for Evaluating Learning Outcomes*. Ithaca, NY: Cornell Lab of Ornithology. Accessed 25 March 2019 at: www.birds.cornell.edu/ [T].

Ferrini-Mundy, J. 2013. Driven by diversity. *Science*, 340:278.

FHC (Fossil Horses in Cyberspace). 2017. Accessed 14 November 2017 at: www.floridamuseum.ufl.edu/fhc.

Florida Museum of Natural History (FLMNH). 2018. Accessed 22 July 2018 at: www.floridamuseum.ufl.edu.

FOSSIL. 2017. FOSSIL Project. Accessed 1 June 2017 at: www.myfossil.org.

Fox, A. & J. Brainard. 2019. University of California takes a stand on open access. *Science*. Accessed 18 March 2019 at: www.sciencemag.org [T].

Fox, S. & M. Griffith. 2007. Hobbyists online. Accessed 8 July 2018 at: www.pewinternet.org [T].

Franceschin, T. 2016. Completion rates are the greatest challenge for MOOCs. Accessed 29 April 2018 at: http://edu4.me/en/completion-rates-are-the-greatest-challenge-for-moocs.

Frechtling, J., M. M. Mark, D. J. Rog, et al. 2010. *The 2010 User-Friendly Handbook for Project Evaluation*. Arlington, VA, National Science Foundation.

Frodeman, R., J. Britt Holbrook, P. S. Bourexis, et al. 2013. Broader Impacts 2.0: seeing – and seizing – the opportunity. *Bioscience*, 63:153–154.

Froschauer, L. 2016. *Bringing STEM to the Elementary Classroom*. Arlington, VA. National Science Teacher Association.

Froschauer, L. 2018. Citizen science. *Science & Children*, 55(8):5.

Frueh, S. 2017. GE crops: a fresh look at the evidence. *National Academies In Focus*, 16:17–19.

Funk, C. & B. Kennedy. 2016. The politics of climate. Pew Research Center. Accessed 12 November 2017 at: www.pewinternet.org [T].

Funk, C. & L. Rainie, 2015. Public and scientists' views on science and society. Pew Research Center. Accessed 15 May 2017 at: www.pewinternet.org [T].

GABI RET (Great American Biotic Interchange Research Experiences for Teachers). 2017. Accessed 15 July 2017 at: http://gabiret.com.

Galetti, M. & R. Costa-Pereira. 2017. Scientists need social media influencers. *Science*, 357:880.

Garet, M. S., A. C. Porter, L. Desimone, B. F. Birman, & K. S. Yoon. 2001. What makes professional development effective? Results from a national sample of teachers. *National Educational Research Association*, 38:915–945.

Gay, T. 2005. *The Physics of Football*. New York: HarperCollins.

GBIF (Global Biodiversity Information Facility). 2015. GBIF releases 2014 science review. Accessed 27 June 2018 at: www.gbif.org [T].

GBIF (Global Biodiversity Information Facility). 2018. Accessed 12 November 2017 at: www.gbif.org.

GeoScienceWorld. 2017. Geology. Accessed 4 September 2017 at: https://pubs .geoscienceworld.org/geology.

Gewin, V. 2013. Free-range learning. *Nature*, 493:441.

Goforth, C. L. 2018. Do children make good citizen scientists? Connected Science Learning. Accessed 13 December 2018 at: http://csl.nsta.org/?s = goforth.

Golden Gate Bridge. 2018. Accessed 9 June 2018 at: http://goldengatebridge.org /visitors.

Grammarist. 2018. Exhibit versus exhibition. Accessed 9 June 2018 at: http://gra mmarist.com/usage/exhibit-exhibition.

Grant, C., B. J. MacFadden, P. Antonenko & V. Perez. 2017. 3D fossils for K–12 education: a case example using the giant extinct shark *Carcharocles megalodon*. *Paleontological Society Papers*, DOI:10.1017/scs.2017.15.

Grants.gov. 2017. Grant making agencies. Accessed 28 November 2017 at: www .grants.gov/web/grants/learn-grants [T].

Gravem, S. A., S. M. Bachhuber, H. K. Fulton-Bennett, et al. 2017. Transformative research is not easily predicted. *Trends in Ecology & Evolution*, 32:825–834.

Great Smoky Mountain Institute. 2018. Tremont partners with Cornell Lab of Ornithology for educator training. Accessed 27 June 2018 at: http://gsmit.org /birdsleuth-cornell-educators.

Greenbaum, D. & M. Gerstein. 2018. What's next for humanity? *Science*, 362:648.

Grobman, A. 1955. Florida State Museum Report [1954–1955] of the Director, University of Florida, Gainesville, November 1955. FLMNH Archives, Office of the Director.

Gross-Loh, C. 2016. Should colleges really eliminate the college lecture? *Atlantic Monthly*, accessed 2 September 2017 at: www.theatlantic.com [T].

GSA (Geological Society of America). 2017. GSA meetings rise to the top. Accessed 20 December 2017 at: www.geosociety.org [T].

Guerrini, C. J., M. A. Majumder, M. J. Lewellyn, & A. L. McGuire. 2018. Citizen science, public policy. *Science*, 361:134–136.

Gunderman, R. 2013. Is the lecture dead? *The Atlantic*. Accessed 3 September 2017 at: www.theatlantic.com/health [T].

Halford, R. 2016. Teaching social media to scientists. *C&EN* 94 (43):21–22. Accessed 22 September 2018 at: https://cen.acs.org [T].

Hall, N. 2014. The Kardashian index: a measure of discrepant social media profile for scientists. *Genome Biology*, 15:424, DOI:10.1186/s13059-014-0424-0.

Hand, E. 2011. NSF takes a broad look at broader impacts. *Nature*. Newsblog. Accessed 31 December 2018 at: http://blogs.nature.com/news/2011/12/nsf_t akes_broad_look_at_broade.html.

Handelsman, J., N. Cantor, M. Carnes, et al. 2005. More women in science. *Science*, 309:1190–1191.

Hansen, K. 2012. Communication: say this not that – word choice matters when communicating climate science. *Earth*, 57(2):53.

Hasiuk, F. 2014. Making things geological: 3-D printing in the geosciences. *GSA Today*, 24(8):28–29.

Heckman, J. J. & T. Kautz. 2012. Hard evidence on soft skills. *Labour Economics*, 19:451–464, DOI:10.1016/j.labeco.2012.05.014.

Hiltzik, M. 2018. In UC's battle with the world's largest scientific publisher, the future of information is at stake. *Los Angeles Times*. Accessed 18 December 2018 at: www.latimes.com.

Hirsch, J. E. 2005. An index to quantify an individual's scientific research output. *Proceedings of the National Academy of Sciences*, 102:16569–16572, DOI:10.1073/pnas.0507655102.

Hoganson, J. W. 2011. Skeletons of the flying reptile *Pteranodon* added to the Bismarck Airport fossil exhibit. *Geo News*, 38(7):1–2.

Holt, R. 2018. A tale of two cultures. *Science*, 359:371.

Homes for the Homeless. 2017. Accessed 18 December 2017 at: www.hfhnyc.org.

Hone, K. S. & G. R. El Said. 2016. Exploring the factors affecting MOOC retention: a survey study. *Computers & Education*, 98:157–168.

Honey, M., G. Pearson, & H. Schweingruber. 2014. *STEM Integration in K–12 Education: Status, Prospects, and an Agenda for Research*. Washington, DC, National Academies Press.

Hoppeler, H. H. 2014. The intricacies of characterizing a scientific journal's performance. *Journal of Experimental Biology*, 217:3773.

Hrabowski, F. A. 2011. Boosting minorities in science. *Science*, 331:125.

Hsiang, S., R. Kopp, A. Jima, et al. 2017. Estimating economic damage from climate change in the United States. *Science*, 356:1362–1368.

Hubenthal, M. & J. Judge. 2013. Taking research experiences for undergraduates online. *EOS, Transactions, American Geophysical Union* 94 (17):157–164.

Hulbert, R. C., Jr. 2016. Montbrook site. Accessed 9 June 2018 at: www.floridamuseum.ufl.edu [T].

Hulbert, R. C., Jr., S. C. Wallace, & J. I Bloch. 2018. Oldest Smilodontin (Felidae, Machairodontinae) skull from the late Miocene of North America. Society of Vertebrate Paleontology Abstracts with Programs, 77th Annual Meeting, Calgary, Canada.

iDigBio. 2018. Integrated Digitized Biocollections. Accessed 31 December 2018 at: www.idigbio.org.

iDigFossils. 2017. iDigFossils: engaging K–12 students in integrated STEM via 3D digitization, printing and paleontology. Accessed 22 December 2017 at: www .idigfossils.org.

iDigPaleo. 2017. iDigPaleo: making data and images of millions of insect specimens available on the web. Accessed 7 November 2017 at: www.idigpaleo.org.

Imaged Reality. 2017. 3D Gaia: the virtual reality application for geological field trips. Accessed 8 July 2018 at: www.imagedreality.com/3d-gaia-vr-app.

IMDb (In-Memory Database). 1993. *Jurassic Park*. Accessed 9 December 2017 at: www.imdb.com/title/tt0107290.

IMDb (In-Memory Database). 2015. *The Martian*. Accessed 9 December 2017 at: www.imdb.com/title/tt3659388.

IME (Institute of Museum Ethics). 2009. Institute of Museum Ethics open storage. Accessed 9 June 2018 at: www.museumethics.org/2009/09/open-storage.

IMLS (Institute of Museum and Library Services). 2008. *Nine to Nineteen: Youth in Museums and Libraries: –A Practitioner's Guide.* Washington, DC: Institute of Museum and Library Services.

IMLS (Institute of Museum and Library Services). 2018. Accessed 9 June 2018 at: www.imls.gov.

iNaturalist. 2018. Accessed 24 June 2018 at: www.inaturalist.org.

Institute for Clinical Research Education (ICRE). 2018. Why mentoring matters. Accessed 27 February 2018 at: www.icre.pitt.edu [T].

Jacobson, S. K., M. McDuff, & M. C. Monroe. 2015. *Conservation Education and Outreach Techniques.* Oxford: Oxford University Press.

Jansen, T. 2016. The preschool inside a nursing home. *The Atlantic* (20 January). Accessed 18 December 2017 at: www.theatlantic.com/education/archive/201 6/01/the-preschool-inside-a-nursing-home/424827.

Jarvis, M. 2017. New AAAS president emphasizes making the case for science. *Science*, 355:807.

Johns, A. 2017. Natural philosophy, 1200–1800. University of Chicago course syllabus. Accessed 6 May 2017 at: http://home.uchicago.edu/~johns [T].

Johnson, C. N., C. J. Bradshaw, A. Cooper, R. Gillespie, & B. W. Brook. 2013. Rapid megafaunal extinction following human arrival throughout the New World. *Quaternary International*, 308:273–277.

Jones, M. G., E. Corin, T. Andre, G. Childers, & V. Stevens. 2017. Factors contributing to lifelong science learning: amateur astronomers and birders. *Journal of Research in Science Teaching*, 54:412–433.

Jordan, K. 2015. MOOC completion rates: the data. Accessed 29 April 2018 at: www.katyjordan.com [T].

Jordan, R. C., H. L. Ballard, & T. B. Phillips. 2012. Key issues and new approaches for evaluating citizen-science learning outcomes. *Frontiers in Ecology and the Environment*, 10:307–309.

Kaiser, J. 2017a. The preprint dilemma. *Science*, 357:1344–1349.

Kaiser, J. 2017b. Biology preprints take off. *Science*, 358:1523.

Kaufmann, A. & F. Tödtling. 2000. Science–industry interaction in the process of innovation – the importance of boundary-crossing between systems. 40th Congress of the European Regional Science Association: "European Monetary Union and Regional Policy," 29 August to 1 September 2000, Barcelona, Spain, European Regional Science Association (ERSA), Louvain-la-Neuve.

Kays, J. 2018. Communicating science for good. UF Center pioneering tactics for creating change. *Explore*. Accessed 17 March 2019 at: https://spark.adobe.com /page/rj6sR39sonrFO.

Kellogg Foundation. 2006. Logic model development guide. Accessed 31 December 2018 at: www.wkkf.org/resource-directory [T].

Kemple, K. M., J. Oh, E. Kenney, & T. Smith-Bonahue. 2016. The power of outdoor play and play in natural environments. *Childhood Education*, 92:446–454.

Keoun, B. 2000. Replica of dinosaur fossil gives O'Hare passengers monstrous welcome. *Chicago Tribune*. Accessed 25 March 2019 at: http://articles .chicagotribune.com [T].

Kincheloe, J. L. & R. A. Horn (eds.). 2007. *The Praeger Handbook of Education and Psychology: Volume 1*. Westport, CT: Praeger.

Kintisch, E. 2014. Is ResearchGate Facebook for Science? *Science*, DOI:10.1126/ science.caredit.a1400214.

Korf, M. 2015. Communicating Research to Public Audiences (CRPA) Program: collecting, sharing and building from successful practices. Accessed 5 September 2017 at: https://crpaworkshop.wordpress.com.

Korn, R. 1995. An analysis of differences between visitors at natural history museums and science centers. *Curator*, DOI:10.1111/j.2151-6952.1995. tb01051.x.

Korn, R. 2018. *Intentional Practice for Museums: A Guide for Maximizing Impact*. New York: Rowman & Littlefield,.

Kosmala, M., A. Wiggins, A. Swanson, & B. Simmons. 2016. Assessing data quality in citizen science. *Frontiers in Ecology and the Environment*, 14:551–560.

Kowaltowski, A. J. & M. F. Oliveira. 2019. Plan S: unrealistic capped fee structure. *Science*, 363:461.

Krajcik, J. 2015. Three-dimensional instruction. *The Science Teacher*, 53:50–52.

Krajcik, J. & I. Delen 2017. How to support learners in developing usable and lasting knowledge of STEM. *International Journal of Education in Mathematics, Science and Technology*, 5:21–28, DOI:10.18404/ijemst.16863.

Krogue, K. 2013. Great presentations: tips from great presenters. *Forbes*. Accessed 4 September 2017 at: www.forbes.com [T].

Kunen, J. S. 2017. Opening minds behind bars. *Columbia Magazine*, summer: 21–27. Accessed 26 December at: https://magazine.columbia.edu [T].

Kwon, D. 2017. Major German universities cancel Elsevier contracts. *The Scientist*. Accessed 20 September 2017 at: www.the-scientist.com [T].

LaBorie, K. 2015. Virtual events: engaged participants vs. disengaged attendees. Association for Talent Development. Accessed 22 September 2018 at: www .td.org [T].

Landivar, L. C. 2013. Disparities in STEM employment by race, sex, and Hispanic origin. American Community Survey Reports. Accessed 13 May 2017 at: www .census.gov [T].

Lave, J. & E. Wenger 1991. *Situated Learning: Legitimate Peripheral Participation (Learning in Doing: Social, Cognitive and Computational Perspectives)*. Cambridge: Cambridge University Press.

Lavery, J. V. 2018. Building an evidence base for stakeholder engagement. *Science*, 361:554–556.

LDS (Latter Day Saints). 2017. General list of disabilities. Accessed 26 December 2017 at: www.lds.org [T].

Lederman, D. 2018. MOOCs: Fewer new students, but more are paying. Retrieved 22 March 2018 at: www.insidehighered.com [T].

Leeming, J. 2017. Dear Dr. Elena: How outreach kills the science stereotype. *Naturejobs*. Accessed 27 February 2018 at: http://blogs.nature.com/nature jobs/2017/07/05 [T].

Leiserowitz, A., E. Maibach, C. Rosner-Renouf, et al. 2017. Politics & global warming, March 2018. Yale Program on Climate Change Communication.

Leitat. 2018. Communication vs. dissemination: what's the difference? Accessed 27 August 2018 at: https://projects.leitat.org. [T].

Leshner, A. 2010. Scientists and science centers: a great glocal [sic.] partnership opportunity. Talk SA 23 at Association of Science – Technology Centers (ASTC) annual meeting, Honolulu, Hawaii, 2–5 October 2010. Accessed 6 May 2017 at: www.astc.org [T].

Leshner, A. I. 2012. Capably communicating science. *Science*, 337:777.

Levine, A. G. 2014. An explosion of bioinformatic careers. *Science*. Accessed 12 November 2017 at: www.sciencemag.org [T].

Levitt, P. 2015. Museums must attract diverse visitors or risk irrelevance. *The Atlantic*. Accessed 12 December 2017 at: www.theatlantic.com/politics [T].

Lewenstein, B. V. 1992. The meaning of "public understanding of science" in the United States after World War II. *Public Understanding of Science*, 1:45–68.

Lima, S. & F. Linzinger. 2017. Breaking free: how Cincinnati Museum Center chose to be a museum without walls. *Informal Learning Review*, 142:8–13.

Lima-Ribeiro, M. S. & J. A. Felizola Diniz-Filho. 2013. American megafaunal extinctions and human arrival: improved evaluation using a meta-analytical approach. *Quaternary International*, 299:38–52.

Lindner, R. M. 1955. *The Fifty-Minute Hour: A Collection of True Psychoanalytic Tales*. Oxford: Rinehart.

Lloyd, R. L., R. Neilson, S. King, & M. Dyball. 2012. *Review of Informal Science Learning*. London: Welcome Trust.

Lohr, S. 2017. Where the STEM jobs are (and where they aren't). *New York Times*. Accessed 30 November 2017 at: www.nytimes.com [T].

Los Alamos Science Fest. 2018. Accessed 31 December 2018 at: www .losalamossciencefestival.com.

Lucas, F. A. 1927. General guide to the exhibition halls at the American Museum of Natural History. American Museum of Natural History. Accessed 25 March 2019 at: http://hdl.handle.net/2246/6685

Lucibella, M. 2015. National Science Foundation issues open access policy. Accessed 11 November 2018 at: www.aps.org/publications/apsnews/updates/access.cfm.

Lucky, A., A. M. Savage, L. M. Nichols, et al. 2014. Ecologists, educators, and writers collaborate with the public to assess backyard diversity in the School of Ants Project. *Ecosphere*, 5(7):1–23.

Lundgren, L. & K. J. Crippen. 2017. Developing social paleontology: a case study implementing innovative social media applications. In D. Remenyi (ed.), *Social Media Excellence Awards 2017: An Anthology of Case Histories* (pp. 11–26). Reading: Academic Conferences and Publishing International Ltd.

Lynas, M. 2015. Even in 2015, the public doesn't trust scientists. *Washington Post*. Accessed 31 December 2015 at: www.washingtonpost.com [T].

Ma, J. & S. Baum. 2016. Trends in community colleges: enrollment, prices, student debt, and completion. College Board Research Brief. Accessed 13 April 2018 at: https://trends.collegeboard.org/sites/default/files/trends-in-community-colleges-research-brief.pdf.

MacCallum. C. J. & H. Parthasarathy. 2006. Open access increases citation rate. *PLoS Biology*, 4:e176, DOI:10.1371/journal.pbio.0040176.

MacFadden, B. J. 1976. Magnetic polarity stratigraphy of the Chamita Formation stratotype (Mio-Pliocene of north-central New Mexico). Unpublished PhD dissertation, Columbia University.

MacFadden, B. J. 1977. Magnetic polarity stratigraphy of the Chamita Formation stratotype (Mio-Pliocene of north-central New Mexico). *American Journal of Science*, 277:769–800.

MacFadden, B. J. 1989. Dental character variation in paleopopulations and morphospecies of fossil horses and extant analogues. In D. R. Prothero & R. M. Schoch (eds.), *The Evolution of Perissodactyls* (pp. 128–141). New York: Clarendon Press.

MacFadden, B. J. 1992. *Fossil Horses: Systematics, Paleobiology, and Evolution of the Family Equidae*. New York: Cambridge University Press.

MacFadden, B. J. 2007. Morris F. Skinner. Society of Vertebrate Paleontology. Accessed 28 February 2018 at: http://vertpaleo.org/Awards [T].

MacFadden, B. J. 2009. Training the next generation of scientists about broader impacts. *Social Epistemology*, 23:239–248.

MacFadden, B. J. 2017. Vertebrate paleontology at the Florida Museum of Natural History, University of Florida: the past 60 years of research and education. *Bulletin of the Florida Museum of Natural History*, 55:51–87.

MacFadden, B. J. & R. P. Guralnick. 2016. Horses in the cloud: big data exploration and mining of fossil and extant *Equus* (Mammalia: Equidae). *Paleobiology*, 43:1–14, DOI:10.1017/pab.2016.42.

MacFadden, B. J., D. A. Dunckel, S. Ellis, et al. 2007. Natural history museum visitors' understanding of evolution. *Bioscience*, 57:875–882.

MacFadden, B. J., L. H. Oviedo, G. Seymour, & S. Ellis. 2012. Fossil horses, orthogenesis, and communicating evolution in museums. *Evolution: Education and Outreach*, DOI:10.1007/s 12052-012-0394-1

MacFadden, B. J., J. I. Bloch, H. Evans, et al. 2014. Temporal calibration and biochronology of the Centenario Fauna, early Miocene of Panama. *Journal of Geology*, 122:113–135.

MacFadden, B. J., L. Lundgren, B. A. Dunckel, S. Ellis, & K. Crippen. 2016. Amateur paleontological societies and fossil clubs, interactions with professional paleontologists, and the rise of 21st Century social paleontology in the United States. *Paleontologica Electronica*, 19(2):1–19.

Mall of America. 2018. Accessed 9 June 2018 at: www.mallofamerica.com.

Mardis, M. A., E. S. Hoffman, & F. P. Martin. 2012. Toward Broader Impacts: making sense of NSF's Merit Review criteria in the context of the National

Science Digital Library. *Journal of the American Society for Information Science and Technology*, 63:1758–1772.

Markman, A. 2017. Your team is brainstorming all wrong. *Harvard Business Review*. Accessed 31 December 2018 at: https://hbr.org.

Marshall, C. R., S. Finnegan, E. C. Clites, et al. 2018. Quantifying the dark data in museum fossil collections as palaeontology undergoes a second digital revolution. *Biology Letters*, 14(9):20180431.

Martens, M. L. & M. L. Carvalho. 2017. Key factors of sustainability in project management context: a survey exploring the project managers' perspective. *International Journal of Project Management*, 35:1084–1102, DOI:10.1016/j.ijproman.2016.04.004.

Martin, P. S. 1984. Prehistoric overkill: the global model. In P. S. Martin & R. G. Klein (eds.), *Quaternary Extinctions: A Prehistoric Revolution* (pp. 354–403). Tucson, AZ: University of Arizona Press.

Marxentlabs. 2015. What is virtual reality? Accessed 8 July 2018 at: www.marxentlabs.com [T].

Maxwell, C. 2016. What blended learning is – and isn't. Accessed 7 July 2018 at: www.blendedlearning.org [T].

Mazuzan, G. T. 1994. The National Science Foundation: a brief history. Accessed 7 May 2017 at: www.nsf.gov [T].

McCauley, D. J. 2017. Digital nature: are field trips a thing of the past? *Science*, 359:298–300.

McComas, W. F. & N. Nouri. 2016. The nature of science and the *Next Generation Science Standards*: analysis and critique. *Journal of Science Teacher Education*, 27:555–576.

McCullough, D. 1977. *The Path Between the Seas*. New York: Simon & Schuster.

McGinnis, P. 2017. STEM integration: a tall order. *Science Scope*, 41(1):1.

McKenzie, L. 2018. Heavyweight showdown over research access. *Inside Higher Ed*. Accessed 18 March 2019 at: www.insidehighered.com [T].

McNutt, M. 2013. Leveling the playing field. *Science*, 341:317.

McNutt, M., M. Bradford, J. M. Drazen, et al. 2018. Transparency in authors' contributions and responsibilities to promote integrity of scientific publication. *Proceedings National Academy of Sciences*, DOI:10.1073/pnas.1715374115.

McPhee, J. 1981. *Basin and Range*. New York: Farrar, Straus & Giroux.

Merriam-Webster. (2017). [Dictionary]. Last accessed 1 July 2018 at: www.merriam-webster.com.

Mervis, J. 2007. Pilot NSF program flies into stiff community headwinds. *Science*, 318:1365–1366.

Mervis, J. 2009. NSF boosts success rates, but at what price. *Science*, 326:1181–1182.

Mervis, J. 2010. From the outside looking in. *Science*, 329:270–273.

Mervis, J. 2014. Congress, NSF spar on access to grant files. *Science*, 346:152–153.

Mervis, J. 2016. NSF director unveils big ideas, with eye on the next president and Congress. Accessed 12 November 2018 at: www.sciencemag.org [T].

Mervis, J. 2017a. The sequel: influential House member plans to rekindle debate over NSF policies. *Science*, DOI:10.1126/science.aal0916.

Mervis, J. 2017b. Polarizing head of House science panel to retire. *Science*, 358:708.

Metz, S. 2017. Smartphones: challenge or opportunity? *Science Teacher*, 84(9):6.

Metz, S. 2018. Phenomenal science. *Science Teacher*, 85(3):6.

Microsoft. 2018. Present your data in a Gantt chart in Excel. Accessed 18 November 2018 at: https://support.office.com/en-us [T].

Miller, J. D., E. C. Scott, & S. Okamoto. 2006. Public acceptance of evolution. *Science*, 313:765–766.

Mills, K., E. Seligman, & D. J. Ketelhut. 2017. Using apps that support scientific practices. *Science Teacher*, 84 (9)14–17.

Mills, M. 1988. Guardian dragon will doze – "Golden Fleece" award is about to go into the attic. *Toledo Blade*, 16 October. Accessed 9 June 2017 at: https://news .google.com/newspapers?id = ajFPAAAAIBAJ&sjid = -QIEAAAAIBAJ&p g = 6796,927271&dq = golden-fleece-award&hl = en.

MIT (Massachusetts Institute of Technology). 2018. OpenCourseWare. Accessed 14 April 2014 at: https://ocw.mit.edu.

Monbiot, G. 2018. Scientific publishing is a rip-off: we fund the research – it should be free. *Guardian*. Accessed 1 December 2018 at: www.theguardian.com.

Moreno, N. 2005. Science education partnerships: being realistic about meeting expectations. *Cell Biology Education*, 4:30–32.

Morphosource. 2018. Accessed 22 July 2018 at: www.morphosource.org.

NABG (National Association of Black Geoscientists). 2018. Accessed 27 February 2018 at: www.nabg-us.org.

NABI (National Alliance for Broader Impacts). 2017. Accessed 16 June 2017 at: https://broaderimpacts.net.

NACME (National Action Council for Minorities in Engineering). 2017. Underrepresented minorities in STEM. Accessed 21 December 2017 at: www .nacme.org/underrepresented-minorities.

NAP (National Academies Press). 2012. Education for life and work. Developing transferable knowledge and skills in the 21st Century. Accessed 9 April 2018 at: www.nap.edu [T].

NAS (National Academies of Sciences, Engineering, and Medicine). 1990. *Volunteers in Public Schools*. Washington, DC: National Academies Press.

NAS (National Academies of Sciences, Engineering, and Medicine). 1997. *Advisor, Teacher, Role Model, Friend: On Being a Mentor to Students in Science and Engineering*. Washington, DC: National Academies Press.

NAS (National Academies of Sciences, Engineering, and Medicine). 2009. *On Being a Scientist*. Washington, DC: National Academies Press.

NAS (National Academies of Sciences, Engineering, and Medicine). 2012. *A Framework for K–12 Science Education*. Washington, DC: National Academies Press. Accessed 4 April 2018 at: www.nap.edu [T].

NAS (National Academies of Sciences, Engineering, and Medicine). 2014. *STEM Integration in K–12 Education*. Washington, DC: National Academies Press. Accessed 31 December 2018 at: www.nap.edu [T].

NAS (National Academies of Sciences, Engineering, and Medicine). 2017a. *Advancing Concepts and Models for Measuring Innovation: Proceedings of a Workshop*. Washington, DC: National Academies Press.

NAS (National Academies of Sciences, Engineering, and Medicine). 2017b. *Communicating Science Effectively: A Research Agenda*. Washington, DC: National Academies Press. Accessed 4 September 2017 at: www.nap.edu [T]

NASA (National Aeronautics and Space Administration). 2018a. John F. Kennedy moon speech. Accessed 16 September 2018 at: https://er.jsc.nasa.gov/seh/ric etalk.htm

NASA (National Aeronautics and Space Administration). 2018b. The first person on the moon. Accessed 16 September 2018 at: www.nasa.gov [T].

National Audubon Society. 2017. Accessed 7 November 2017 at: www.audubon.org.

National Eye Institute. 2018. Accessed 23 August 2018 at: https://nei.nih.gov/ health/color_blindness/facts_about

National Postdoctoral Association. 2018. Accessed 27 February 2018 at: www.nationalpostdoc.org.

Nature. 2017. Transferable skills: STEM must branch out. *Nature*, DOI:10.1038/ nj7644-277b.

NCAA (National Collegiate Athletics Association), 2018. Football attendance 2017, accessed 31 December 2018 at: http://fs.ncaa.org/Docs/stats/football_records/ Attendance/2017.pdf

NCBYS (National Center for Blind Youth in Science). 2017. Dr. Geerat J. Vermeij Biology. Accessed 26 December 2017 at: www.blindscience.org/vermeij.

NCEE (National Center for Education and the Economy. 2017. Are teachers valued by society? Accessed 24 May 2017 at: www.ncee.org/wp-content/uploads/201 6/04/SocietyValueTeachersFINAL-1.pdf.

NCES (National Center for Educational Statistics). 2017. Accessed 12 December 2017 at: https://nces.ed.gov.

NCLB. 2001. No Child Left Behind Act of 2001. Accessed 4 April 2018 at: www .congress.gov [T].

NCSL (National Conference of State Legislators). 2017.Tips for making effective PowerPoint presentations. Accessed 28 August 2018 at: www.ncsl.org [T].

NEA (National Education Association). 2017. Global competence is a 21st century imperative. Accessed 25 March 2019 at: https://multilingual .madison.k12.wi.us/files/esl/NEA-Global-Competence-Brief.pdf.

New York Times (The Editorial Board). 2017. Let prisoners learn while they serve. *New York Times*. Accessed 18 December 2017 at: www.nytimes.com [T].

New York Times. 2018. [Mobile app]. Accessed 8 July 2018 at: www.nytimes.com [T].

NGSS (Next Generation Science Standards). 2013. Accessed 15 July 2017 at: www.nextgenscience.org.

NHM Alive. 2018. David Attenborough's Natural History Museum Alive. Accessed 7 July 2018 at: www.naturalhistorymuseumalive.com.

Nichols, L., L. Brako, S. M. Rivera, et al. 2017. What do revised U.S. rules mean for human research? *Science*, 357:650–651.

Nisbet, M. C. & D. A. Scheufele. 2009. What's next for science communication? Promising directions and lingering distractions. *American Journal of Botany*, 96 (10):1767–1778.

Noack, R. & L. Gamio. 2015. The world's languages, in 7 maps and charts. *Washington Post*. Accessed 24 May 2017 at: www.washingtonpost.com [T].

Nokes, Sebastian. 2007. *The Definitive Guide to Project Management*, 2nd edition. London: Financial Times/Prentice Hall.

Noonan, R. 2017. Women in STEM: 2017 Update. Accessed 27 March 2019 at: www .commerce.gov/sites/default/files/migrated/reports/women-in-stem-2017-update.pdf.

Normile, D. 2018. China narrows U.S. lead in R&D spending. *Science*, 362:276.

NPS (National Park Service). 1996. *Volunteers in Parks*. Washington, DC: National Park Service (brochure, not paginated).

NPS (National Park Service). 2017. Volunteer with us. Accessed 14 June 2018 at: www.nps.gov [T].

NRC (National Research Council). 1990. *Volunteers in Public Schools*. Washington, DC: National Academies Press.

NRC (National Research Council). 1996. *National Science Education Standards*. Washington, DC: National Academy Press.

NSB (National Science Board). 2016. Science and engineering indicators 2016. Accessed 22 December 2017 at: www.nsf.gov/statistics/2016/nsb20161/#.

NSF (National Science Foundation). 1951. The first annual report of the National Science Foundation 1950–51. Accessed 8 June 2017 at: www.nsf.gov/about/hi story/ann_report_first.pdf.

NSF (National Science Foundation). 1951–2015. NSF budget history by account from FY 1951 to 2015. Accessed 9 June 2017 at: https://catalog.data.gov/data set/nsf-budget-history-by-account-from-fy–1951.

NSF (National Science Foundation). 1952. The second annual report of the National Science Foundation 1952. Accessed 8 June 2017 at: www.nsf.gov/pubs/1952/ annualreports/start.htm.

NSF (National Science Foundation). 1956. The sixth annual report of the National Science Foundation 1956. Accessed 12 November 2017 at: www.nsf.gov/pubs/ 1956/annualreports/start.htm.

NSF (National Science Foundation). 1957. The seventh annual report of the National Science Foundation 1957. Accessed 12 November 2017 at: www .nsf.gov/pubs/1957/annualreports/start.htm.

NSF (National Science Foundation). 1958. The eighth annual report of the National Science Foundation 1958., Accessed 12 November 2017 at: www.nsf.gov/pubs/ 1958/annualreports/start.htm.

NSF (National Science Foundation). 1962. The twelfth annual report of the National Science Foundation 1962. Accessed 9 June 2017 at: www.nsf.gov/pubs/1962/ annualreports/start.htm.

NSF (National Science Foundation). 1974. National Science Foundation annual report 1974. Accessed 9 June 2017 at: http://files.eric.ed.gov/fulltext/E D115473.pdf.

NSF (National Science Foundation). 2003. Communicating research to public audiences. Program Solicitation 03-509. Accessed 29 August 2018 at: www .nsf.gov/pubs/2003/nsf03509/nsf03509.html.

NSF (National Science Foundation). 2007. (Dear Colleague Letter) Broader Impacts Review Criterion. Accessed 9 June 2017 at: www.nsf.gov/pubs/2007/nsf07046/ nsf07046.jsp.

NSF (National Science Foundation). 2008. Broadening Participation at the National Science Foundation: a framework for action. Accessed 18 December 2018 at: www.nsf.gov/od/broadeningparticipation/nsf_ frameworkforaction_0808.pdf

NSF (National Science Foundation). 2009a. Partnerships for International Research and Education (PIRE) Program Solicitation 09–505; accessed 8 July 2017 at: www.nsf.gov/pubs/2009/nsf09505/nsf09505.htm.

NSF (National Science Foundation). 2009b. FY 2009 performance and financial highlights. Accessed 11 November 2017 at: www.nsf.gov/pubs/2010/ nsf10002/nsf10002.pdf.

NSF (National Science Foundation). 2009c. Dear Colleague Letter: Climate change education (NSF 09-058). Accessed 14 November 2017 at: www.nsf.gov/pubs/ 2009/nsf09058/nsf09058.jsp.

NSF (National Science Foundation). 2009d. Proposal and Award Policies and Procedures Guide, April 2009 (NSF 09–058). Accessed 27 February 2018 at: www.nsf.gov/publications/pub_summ.jsp?ods_key = nsf0929.

NSF (National Science Foundation). 2009e. NSF graduate stem fellows in K–12 education (GK–12). Accessed 9 April 2018 at: www.nsf.gov/pubs/2009/ nsf09549/nsf09549.htm.

NSF (National Science Foundation). 2010a. Responsible conduct of research (RCR): frequently asked questions. Accessed 7 November 2017 at: www.nsf.gov/pubs/ policydocs/rcr/faqs_mar10.pdf.

NSF (National Science Foundation). 2010b. FY 2010 performance and financial highlights. Accessed 14 November 2017 at: www.nsf.gov/publications/pub_ summ.jsp?ods_key = nsf11002.

NSF (National Science Foundation). 2010c. Geoscience education (GeoEd). Accessed 9 April 2018 at: www.nsf.gov/publications/pub_summ.jsp? ods_key = nsf10512.

NSF (National Science Foundation). 2011. Dear Colleague Letter: Partnerships for International Research and Education (PIRE) with Science, Engineering, and Education for Sustainability (SEES) Focus (NSF 11–205). Accessed 12 November 2017 at: www.nsf.gov/pubs/2011/nsf11025/nsf11025.jsp.

NSF (National Science Foundation). 2012a. Proposal & Award Policies & Procedures Guide (PAPPG, document 13-1). Accessed 15 June 2017 at: www .nsf.gov/pubs/policydocs/pappguide/nsf13001/nsf13_1.pdf.

NSF (National Science Foundation). 2012b. Dear Colleague Letter: Research Experiences for Teachers (RET): funding opportunity in the biological sciences. Accessed 9 April 2018 at: www.nsf.gov/pubs/2012/nsf12075/ns f12075.jsp.

NSF (National Science Foundation). 2014a. Science and Technology Centers: integrative partnerships. Program Solicitation 14-600. Accessed 8 July 2017 at: www.nsf.gov/pubs/2014/nsf14600/nsf14600.htm.

NSF (National Science Foundation). 2014b. How many undergraduates enroll in community colleges? Science and engineering indicators. Accessed 13 April 2018 at: www.nsf.gov/nsb/sei/edTool/data/college-04.html.

NSF (National Science Foundation). 2015. Big Data @ NSF. Accessed 12 November 2017 at: www.nsf.gov/cise/bigdata.

NSF (National Science Foundation). 2016a. Consolidated Appropriations Act of FY 2016. Accessed 9 June 2017 at: www.nsf.gov/about/congress/114/highlights/cu16_0104.jsp.

NSF (National Science Foundation). 2016b. Proposal & Award Policies & Procedures Guide (17-1, PAPPG). Accessed 16 June 2017 at: www.nsf.gov/pubs/policydocs/pappg17_1/nsf17_1.pdf.

NSF (National Science Foundation). 2017a. A timeline of NSF history. Accessed 8 June 2017 at: www.nsf.gov/about/history/overview-50.jsp.

NSF (National Science Foundation). 2017b. NSF & Congress. Accessed 31 December 2018 at: www.nsf.gov/about/congress/115/highlights/cu17_0508.jsp.

NSF (National Science Foundation). 2017c. EPSCoR. Accessed 16 June 2017 at: www.nsf.gov/od/oia/programs/epscor/Eligibility_Tables/FY2017_Eligibility.pdf.

NSF (National Science Foundation). 2017d. Small Business Innovation Research Program Phase I (SBIR). Program Solicitation 17-544. Accessed 9 July 2017 at: www.nsf.gov/pubs/2017/nsf17544/nsf17544.htm.

NSF (National Science Foundation). 2017e. Research Experiences for Undergraduates (REU) Solicitation. Accessed 15 July 2017 at: www.nsf.gov/pubs/2013/nsf13542/nsf13542.pdf.

NSF (National Science Foundation). 2017f. Dear Colleague Letter: Public participation in science, technology, engineering, and mathematics research: capacity-building, community-building, and direction-setting. Accessed 15 July 2017 at: www.nsf.gov/pubs/2017/nsf17047/nsf17047.jsp.

NSF (National Science Foundation). 2017g. Innovative Technology Experiences for Students and Teachers (ITEST) Solicitation 15-599. Accessed 15 July 2017 at: www.nsf.gov/pubs/2015/nsf15599/nsf15599.htm#toc.

NSF (National Science Foundation). 2017h. NSF, science communication, and you. Accessed 4 September 2017 at: www.nsf.gov/about/congress/reports/NSFScienceCommunicationAndYou.pdf.

NSF (National Science Foundation). 2017i. Research Coordination Networks (RCN). Accessed 7 November 2017 at: www.nsf.gov/pubs/2017/nsf17594/nsf17594.htm.

NSF (National Science Foundation). 2017j. Advancing Digitization of Biodiversity Collections (ADBC). Accessed 7 November 2017 at: www.nsf.gov/funding/pgm_summ.jsp?pims_id = 503559.

NSF (National Science Foundation). 2017k. Major Research Instrumentation (MRI) Program: instrument acquisition or development. Program Solicitation 18-513. Accessed 12 November 2017 at: www.nsf.gov/pubs/2018/nsf18513/nsf18513.htm.

NSF (National Science Foundation). 2017l. Dear Colleague Letter (DCL): Enabling new collaborations between computer and information science & engineering (CISE) and social, behavioral and economic sciences (SBE) research communities (NSF 17-019). Accessed 15 November 2017 at: www.nsf.gov/pubs/2017/nsf17019/nsf17019.jsp.

NSF (National Science Foundation). 2017m. Building community and capacity in data intensive research in education (BCC-EHR). Program solicitation NSF 17-532. Accessed 14 November 2017 at: www.nsf.gov/pubs/2017/nsf17532/nsf17532.htm.

NSF (National Science Foundation). 2017n. NSF INCLUDES. Accessed 14 November 2017 at: www.nsf.gov/news/special_reports/nsfincludes/index.jsp.

NSF (National Science Foundation). 2017o. Science, Engineering and Education for Sustainability NSF-wide investment (SEES): SEES mission statement. Accessed 14 November 2017 at: www.nsf.gov/funding/pgm_summ.jsp?pims_id = 504707.

NSF (National Science Foundation). 2017p. 10 big ideas for future NSF investments. Accessed 14 November 2017 at: www.nsf.gov/about/congress/reports/nsf_big_ideas.pdf.

NSF (National Science Foundation). 2017q. *Women, Minorities, and Persons with Disabilities in Science and Engineering: 2017*. Special Report NSF 17-310. Arlington, VA: NSF. Accessed 27 March 2019 at: www.nsf.gov/statistics/wmpd.

NSF (National Science Foundation). 2017r. ADVANCE: increasing the participation and advancement of women in academic science and engineering careers (ADVANCE). Accessed 26 December 2017 at: www.nsf.gov/funding/pgm_summ.jsp?pims_id = 5383.

NSF (National Science Foundation). 2017s. EAR Education and Human Resources (EH). Accessed 26 December 2017 at: www.nsf.gov/funding/pgm_summ.jsp?pims_id = 13414.

NSF (National Science Foundation) 2017t. Discovery Research PreK–12 (DRK–12). Program Solicitation 17-584. Accessed 4 September 2018 at: www.nsf.gov/pubs/2017/nsf17584/nsf17584.htm.

NSF (National Science Foundation). 2017u. REU site: nanotechnology REU with a focus on community colleges. Award 1757967. Accessed 13 April 2018 at: www.nsf.gov/awardsearch/showAward?AWD_ID = 1757967&HistoricalAwards = false.

NSF (National Science Foundation). 2017v. ATE Advanced Technological Engineering Program (17-568). Accessed 13 April 2018 at: www.nsf.gov/pubs/2017/nsf17568/nsf17568.htm.

NSF (National Science Foundation). 2017w. Robert Noyce Teacher Scholarship Program. Accessed 27 March 2019 at: www.nsf.gov/pubs/2017/nsf17541/nsf17541.htm

NSF (National Science Foundation). 2017x. Advancing Informal STEM Education (AISL). Accessed 9 June 2018 at: https://nsf.gov/pubs/2017/nsf17573/nsf17573.htm.

NSF (National Science Foundation). 2018a. NSF funding rate history web API. Accessed 4 January 2018 at: https://catalog.data.gov/dataset/nsf-funding-rate-history-web-api.

NSF (National Science Foundation). 2018b. FY 2019 budget request to Congress. Accessed 12 November 2018 at: www.nsf.gov/about/budget/fy2019/index.jsp.

NSF (National Science Foundation). 2018c. Dear Colleague Letter: Announcement of an effort to expand the INCLUDES National Network (NSF 17-111). Accessed 27 February 2018 at: www.nsf.gov/pubs/2017/nsf17111/nsf17111.jsp.

NSF (National Science Foundation). 2018d. Graduate Research Fellowship Program. Accessed 27 April 2018 at: www.nsfgrfp.org/general_resources/about.

NSF (National Science Foundation). 2018e. Tribal Colleges and Universities Program (TCUP) (NSF 18-546). Accessed 13 April 2018 at: www.nsf.gov/publications/pub_summ.jsp?WT.z_pims_id = 5483&ods_key = nsf18546.

NSF (National Science Foundation). 2018f. Transform your ideas, impact your world. CCIC (Community College Innovation Program). Accessed 13 April 2018 at: www.nsf.gov/news/special_reports/communitycollege.

NSF (National Science Foundation). 2018g. National Science Foundation Research Traineeship Program (NRT). Program Solicitation 18-507. Accessed 30 April 2018 at: www.nsf.gov/pubs/2018/nsf18507/nsf18507.pdf.

NSF (National Science Foundation). 2018h. Proposal & Award Policies & Procedures Guide. Program Solicitation 19-1. Accessed 1 January 2019 at: www.nsf.gov/publications/pub_summ.jsp?ods_key = nsf19001.

NSF (National Science Foundation). 2018i. Broadening Participation. Accessed 12 November 2018 at: www.nsf.gov/od/broadeningparticipation/bp.jsp.

NSF (National Science Foundation). 2018j. Broadening Participation Portfolio. Accessed 12 November 2018 at: www.nsf.gov/od/broadeningparticipation/bp_portfolio_dynamic.jsp.

NSTA (National Science Teaching Association). 2018a. About the Next Generation Science Standards. Accessed 4 April 2018 at: https://ngss.nsta.org [T].

NSTA (National Science Teaching Association). 2018b. NSTA position statement: the nature of science. Accessed 9 June 2018 at: www.nsta.org [T].

NYHS (New York Hall of Science). 2017. Accessed 9 December 2017 at: https://nysci.org.

NYU (New York University). 2017. How to create a research poster: poster basics. Accessed 4 September 2017 at: http://guides.nyu.edu/posters.

Obama, B. 2016. United States health care reform: progress to date and next steps. *Journal of the American Medical Association*, 316(5):525–532, DOI:10.1001/jama.2016.9797.

OECD-Eurostat. 2005. *Oslo Manual: Guidelines for Collecting and Interpreting Innovation Data*, 3rd edition. Paris: OECD. Accessed 8 July 2017 at: www.oecd.org [T].

ORCID. 2017. ORCID connecting researchers and researchers. Accessed 7 November 2017 at: https://orcid.org.

O'Reilly, T. 2005. What is Web 2.0: design patterns and business models for the next generation of software. Accessed 12 November 2017 at: www.oreilly.com [T].

Orzel, C. 2015. What has quantum mechanics ever done for us? *Forbes*. Accessed 31 December 2018 at: www.forbes.com/sites/chadorzel/2015/08/13/what-has -quantum-mechanics-ever-done-for-us/#19f7efc24046

OSU. 2018. Free-choice learning. Oregon State University. Accessed 9 June 2018 at: https://education.oregonstate.edu/free-choice-learning.

Overman, S. 2018. Fighting the stress of teaching to the test. National Education Association. Accessed 4 April 2018 at: www.nea.org/tools/fighting-stress-teaching-to-Test.html.

oVert. 2017. oVert: Open Exploration of Vertebrate Diversity in 3D. Accessed 7 November 2017 at: www.idigbio.org/wiki/index.php/OVert: _Open_Exploration_of_Vertebrate_Diversity_in_3D.

Oxford Living Dictionaries. 2019. Accessed 27 March 2019 at: https://en.oxforddictionaries.com [T].

P21. 2017. Partnership for 21st century learning. Accessed 16 May 2017 at: www.p21.org.

Page, L. M., B. J. MacFadden, J. A. Fortes, P. S. Soltis, & G. Riccardi. 2015. Digitization of biodiversity collections reveals biggest data on biodiversity. *Bioscience*, DOI:10.1093/biosci/biv104.

PaleoTEACH. 2018. Accessed 5 April 2018 at: www.paleoteach.org.

Panama PIRE (Partnerships for International Research and Education). 2017. Panama Canal Project (PCP). Accessed 9 July 2017 at: www.floridamuseum.ufl.edu [T].

Panama PIRE (Partnerships for International Research and Education). 2010. Project management plan. Unpublished document, archived at: www.floridamuseum.ufl.edu [T].

Pandya, R. E. 2012. A framework for engaging diverse communities in citizen science in the U.S. *Frontiers in Ecology & Environment*, 10:314–317, DOI:10.1890/120007.

Passel, J. & D. Cohn. 2008. U.S. population projections: 2005–2050. Pew Research Center. Accessed 11 May 2017 at: www.pewsocialtrends.org [T].

Pathmanathan, S. 2014. Dance me to the end of time and space. Accessed 27 August 2018 at: www.space.com [T].

Paulus, P. B. & B. J. Nijstad (eds.). 2003. *Group Creativity: Innovation Through Collaboration*. Oxford: Oxford University Press.

PBDB (Paleobiology DataBase). 2018. Accessed 12 November 2017 at: https://paleobiodb.org/#.

PBS (Public Broadcasting Service). 2017. Can students return a billion oysters to a New York Harbor? Accessed 23 June 2018 at: www.pbs.org [T].

Pearce, J. M. 2016. Return on investment for open source scientific hardware development. *Science Public Policy*, 43:192–195, DOI:10.1093/scipol/scv034.

PeerJ. 2017. Accessed 20 September 2017 at: https://peerj.com.

Peningian. 1967. *Yearbook*. New York: Port Chester High School.

Pérez-Peña, R. 2012. Top universities test the online appeal of free. *New York Times*. Accessed 27 April 2018 at: www.nytimes.com [T].

Perot Museum of Nature and Science. 2018. Accessed 9 June 2018 at: www.perotmuseum.org.

Persson, O. & W. Glänzel. 2013. Discouraging honorific authorship. *Scientimetrics*, DOI:10.1007/s11192-013-1042-4

Pew Research Center. 2017. Mobile fact sheet. Accessed 8 July 2018 at: www.pewinternet.org [T].

Pielke, R., Jr. 2010. In retrospect: Science – The Endless Frontier. *Nature* 466:922.

Pimiento, C. & M. Balk. 2015. Body-size trends of the extinct giant shark *Carcharocles megalodon*: a deep-time perspective on marine apex predators. *Paleobiology*, 41:479–490, DOI:10.1017/pab.2015.16.

Pimiento C. & C. F. Clements. 2014. When did *Carcharocles megalodon* become extinct? A new analysis of the fossil record. *PLoS ONE*, 9(10): e111086, DOI:10.1371/journal.pone.0111086.

Pimiento, C., D. J. Ehret, B. J. MacFadden, & G. Hubbell. 2010. Ancient nursery area for the extinct giant shark Megalodon from the Miocene of Peru. *PLoS ONE*, 5: e10552.

Pimiento, C., B. J. MacFadden, C. Clements, et al. 2016. Geographic distribution patterns of *Carcharocles megalodon* over time reveal clues about mechanisms of extinction. *Journal of Biogeography*, DOI:10.1111/jbi.12754.

Pinol, N. D. 2010. Science Editor-in-Chief Bruce Alberts wins 2010 Vannevar Bush Award. Accessed 20 July 2018 at: www.aaa.org/news [T].

Pirlo, J., B. J. MacFadden, E. E. Gardner, V. J. Perez, & D. Porcello. 2018. Connecting fossil clubs with K–12 teachers. *Connected Science Learning*. Accessed 23 August 2018 at: http://csl.nsta.org/2018/05/connecting-fossil-clubs.

PMI (Project Management Institute). 2018. Accessed 11 November 2018 at: www.pmi.org.

Pope, J. 2014. What are MOOCs good for?. Accessed 20 April 2018 at: www.technologyreview.com [T].

Porter, A. L., J. Garner, & T. Crowl. 2012. Research coordination networks: evidence of the relationship between funded interdisciplinary networking and scholarly impact. *BioScience*, 62:282–288, DOI:10.1525/bio.2012.62.3.9.

PRB (Population Reference Bureau). 2017. Changing demographics reshape rural America. Accessed 22 December 2017 at: www.prb.org [T].

Princeton University Council on Science and Technology . 2017. The prison and the academy: STEM education and prisoner reentry. Accessed 18 December 2017 at: https://cst.princeton.edu [T].

Prochaska, F. 2018. Basic vs. applied research (from Lawrence Berkeley National Laboratory). Accessed 1 September 2018 at: www.sjsu.edu [T].

Professionals Australia. 2017. Women in STEM in Australia. Accessed 12 May 2017 at: www.professionalsaustralia.org.au [T].

Proxmire, W. 1975–1987. Golden Fleece Awards (Wis Mss 738, Box 158, Folders 1–5). Online facsimile accessed 7 June 2017 at: www.wisconsinhistory.org [T].

Public Law 81-507. 1950. *National Science Foundation Act of 1950*. Accessed 8 June 2017 at: www.nsf.gov [T].

Public Law 92-484. 1972. *Technology Assessment Act of 1972*. Accessed 9 June 2017 at: http://uscode.house.gov [T].

Purington, C. 2017. Designing conference posters. Accessed 2 September 2017 at: http://colinpurrington.com [T].

Pye, G., S. Williams, & L. Dunne. 2016. Student academic mentoring (SAM): peer support and undergraduate study. *Journal of Learning Development in Higher Education*, Special Edition: April:1–23.

Rabesandratana, T. 2018. European funders detail their open-access plan. *Science*, 362:983.

Rader, K. A. & V. E. M. Cain. 2014. *Life on Display*. Chicago, IL: University of Chicago Press.

Rais, P. 2012. From access to inclusion: welcoming the autism community. *Dimensions ASTC (Association of Science – Technology Centers)*, November/December:43–46.

Rajput, A. S. D. 2018. India's Ph.D. scholar outreach requirement. *Science*, 359:1343, DOI:10.1126/science.aat0303.

Ratcliff, J. L. 2018. Community colleges: the history of community colleges, the junior college and the research university, the community college mission. Accessed 27 April 2017 at: http://education.stateuniversity.com [T].

Rea, D. G. & K. M. Nielsen. 2010. A volunteer army for science. *Science*, 329:257.

Reddy, C. 2009. Science citizens. *Science*, 323:1405.

Rees, J. 2016. A brief history of the mobile museum: what it is and what it can be. Unpublished document, University of Kansas. Accessed 12 May 2018 at: https://kuscholarworks.ku.edu [T].

Rehman, J. 2013. The need for critical science journalism. *Guardian*. Accessed 6 December 2017 at: www.theguardian.com [T].

Reich, C. 2012. Changing practices: inclusion of people with disabilities in science museums. *Dimensions ASTC (Association of Science and Technology Centers)*, November/December:22–27.

ResearchGate. 2017. Accessed 8 November 2017 at: www.researchgate.net.

Resmini, M. 2016. The "Leaky Pipeline." *Chemistry*, 22:3522–3534, DOI:10.1002/chem.201600292.

Riben, E., P. Sapienza, & L. Zingales. 2014. How stereotypes impair women's careers in science. *Proceedings of the National Academy of Sciences*, 111:4403–4408, DOI:10.1073/pnas.1314788111.

Rockoff, J. 2004. The impact of individual teachers on student achievement: evidence from panel data. *American Economic Review*, 94:247–252.

Romer, A. S. 1969. Teaching vertebrate paleontology. *Proceedings of the North American Paleontological Convention*, Part A:39–46.

Rosenbaum, J. L. 2008. High-profile journals not worth the trouble. *Science*, 321:1039.

Rothenberg, M. 2010. Making judgements about grant proposals: a brief history of the merit review criteria at the National Science Foundation. *Technology and Innovation*, 12:189–195.

Rudolph, J. L. 2002. *Scientists in the Classroom: The Cold War Reconstruction of American Science Education*, New York: Palgrave.

SACNAS (Society for Advancement of Chicanos/Hispanics and Native Americans in Science). 2018. Accessed 27 February 2018 at: http://sacnas.org.

Santa Fe College. 2018. Santa Fe College (Gainesville, Florida), programs of study. Accessed 13 April 2018 at: www.sfcollege.edu/programs [T].

Santos, G.-P., B. Stoneburg, M. M. Barboza, & I. Magallanes. 2018. Cosplay for science: connecting paleontology, education, and pop culture. *Geological Society of America Abstracts with Programs*, 50(6), DOI:10.1130/abs/2018AM-324738.

Sarewitz, D. 2011. The dubious benefits of broader impact. *Nature*, 472:141.

SCCS (Santa Cruz City Schools). 2018. Two-way immersion DeLaveaga Elementary School. Accessed 2 September 2018 at: http://sccs.net/schools/district_programs/two_way_immersion.

Schaal, B. 2017. Informing policy with science. *Science*, 355:435.

School of Ants. 2018. Accessed 31 December 2018 at: www.schoolofants.org.

Schwalbe, K. 2016. *Information Technology Project Management*, 8th edition. Boston, MA: Cengage Learning.

Science. 2018. A new way to measure innovation. *Science*, 359:721.

Science Friday. 2016. Exploring geologic history with a virtual field trip. Accessed 8 July 2018 at: www.sciencefriday.com [T].

Science Friday. 2017. Accessed 12 December 2017 at: www.sciencefriday.com.

Science Netlinks. 2018. Accessed 7 July 2018 at: www.sciencenetlinks.com.

Scijournal. 2017. Impact factor list. Accessed 12 September 2017 at: www.scijournal.org.

SciStarter. 2018. SciStarter for educators. Accessed 17 June 2018 at: https://scistarter.com/educators.

Screven, C. 1990. Uses of evaluation before, during and after exhibit design. *ILVS (International Laboratory of Visitors Studies) Review* 1(2):36–66.

SEPAL (Science Education Partnership & Assessment Laboratory). 2018. Accessed 27 February 2018 at: www.sfsusepal.org.

Serrell, B. 1998. *Paying Attention: Visitors and Museum Exhibits*. Washington, DC: American Association of Museums.

Serrell, B. 2015. *Exhibit Labels: An Interpretive Approach*, 2nd edition. Lanham, MD: Rowman & Littlefield.

Shirk, J. L., H. L. Ballard, C. C. Wilderman, et al. 2012. Public participation in scientific research: a framework for deliberate design. *Ecology and Society*, DOI:10.5751/ES-04705-170229.

Sills, J. 2018. Quality mentoring. *Science*, 362:22–24.

Silverstein, G. 2008. Using logic models to identify desired impacts and audience objectives. In A. J. Freidman (ed.), *Framework for Evaluating Impacts of Informal Science Education Projects* (pp. 35–40). Arlington, VA; National Science Foundation.

Silvius, G. & J. Tharp (eds.). 2013. *Sustainability Integration for Effective Project Management*. Hershey, PA: IGI Global.

Silvius, G., R. Schipper, & J. Planko. 2012. *Sustainability in Project Management*. Aldershot: Gower Publishing.

Simon, N. 2010. The Participatory Museum. Accessed 23 March 2019 at: www .participatorymuseum.org.

Simon, N. 2011. A radical, simple formula for pop-up museums. Accessed 9 June 2018 at: http://museumtwo.blogspot.com/2011/11/radical-simple-formula-for-pop-up.html.

Skema, C. 2018. Mid-Atlantic Megalopolis: two years in . . .. Talk presented at iDigBio Summit 2018, Gainesville, Florida. Accessed 31 December 2018 at: www.idigbio.org/sites/default/files/workshop-presentations/summit8/Day-1 /15-MAM_Summit_2018.pdf.

Skype a Scientist, 2018. Accessed 17 November 2018 at: www .skypeascientist.com.

Smith, A. A. 2018. Growing number of community colleges focus on diversity and inclusion. *Inside Higher Ed*. Accessed 13 April 2013 at: www .insidehighered.com [T].

SPP (Sustainability in Prisons Project). 2017. Accessed 18 December 2017 at: http://sustainabilityinprisons.org.

Stainsbury, M. 2015. How to calculate the real costs of developing and delivering MOOCs. Accessed 29 April 2018 at: www.ecampusnews.com [T].

Stanley, S. M. 1973. An explanation for Cope's Rule. *Evolution*, 27:1–26, DOI:10.2307/2407115.

Statista. 2017. U.S. print media industry: statistics and facts. Accessed 6 December 2017 at: www.statista.com [T].

Statista. 2018. [Use of mobile devices for Millennials and Generation X]. Accessed 8 July 2018 at: www.statista.com.

STEM Teaching Tools. 2018. Using phenomena in NGSS-designed lessons and units. Accessed 5 April 2018 at: http://stemteachingtools.org/brief/42.

Strange, K. 2008. Authorship: why not just toss a coin? *American Journal of Physiology and Cell Physiology*, 295:C567–C575, DOI: 10.1152/ ajpcell.00208.2008.

Streaming Science. 2018. https://streamingscience.com.

Superville, D. R. 2017. More than 1 million K–12 students are homeless. what one district is doing about it. Accessed 3 September 2018 at: http://blogs .edweek.org [T].

Swift, A. 2017. In US, belief in creationist view of evolution at new low. Gallup. Accessed 24 May 2017 at: www.pewinternet.org [T].

Sykes, K. 2007. The quality of public dialogue. *Science*, 318:1349.

Syms, C. 2016. Why volunteers are your museum's secret weapon. Accessed 14 June 2018 at: https://museumhack.com [T].

TACC (Texas Advanced Computing Center). 2018. Accessed 22 July 2018 at: www .tacc.utexas.edu.

Tanner, K. D. 2000. Evaluation of scientist–teacher partnerships: benefits to scientist participants. National Association of Research in Science Teaching (NARST) Annual Conference, New Orleans, LA.

Tanner, K. D., L. Chatman, & D. E. Allen. 2003. Science teaching and learning across the school–university divide: cultivating conversations through scientists–teacher partnerships. *Cell Biology Education*, 2:195–201.

Tapanila, L. & I. Rahman. 2016. Virtual paleontology. *Paleontological Society Papers*, 22:1–209.

Technion 2017. Average H-index may vary between different disciplines. Accessed 20 September at: https://library.technion,ac.il/en [T].

Tech Target. 2018. Accessed 27 July 2018 at: https://searchsoftwarequality .techtarget.com [T].

TED Talks. 2017. Accessed 12 December 2017 at: www.ted.com/talks.

tes. 2014. KS2 – intro to databases – fun, practical – complete unit. Accessed 8 July 2018 at: www.tes.com [T].

Tomanek, D. 2005. Building successful partnerships between K–12 and universities. *Cell Biology Education*, 4:28–29.

Tovey, C. A. 2017. In defense of basic research. *Science*, 355:804.

Trainor, S. 2015. How community colleges changed the whole idea of education in America. *Time*. Accessed 27 April 2018 at: http://time.com [T].

Trilling, B. & C. Fadel. 2012. *21st Century Skills: Learning for Life in Our Times*. Hoboken, NJ: Jossey-Bass (Wiley).

Tripadvisor. 2018. Things to do in Gainesville. Accessed 9 June 2018 at: www .tripadvisor.com/Attractions-g34242-Activities-Gainesville_Florida.html.

UF IRB. 2017. University of Florida Institutional Review Board IRB-02. UF Campus/ non-medical. Accessed 9 December 2017 at: http://irb.ufl.edu/irb02.html.

UF Teach. 2018. UF Teach program, College of Education. Accessed 27 April 2018 at: https://education.ufl.edu [T].

UIGC (University of Illinois Graduate College). 2018. Writing postdoctoral mentoring plans. Accessed 27 February 2018 at: https://grad.illinois.edu/ postdocs [T].

UNESCO. 2010. UNESCO teaching and learning for a sustainable future. Accessed 9 July 2018 at: www.unesco.org [T].

United Nations. 2017. Framework Convention on Climate Change: Paris Agreement – status of ratification. Accessed 6 May 2017 at: http://unfccc.int/ paris_agreement/items/9444.php.

University of Washington, College of the Environment. 2019. https://environment
.uw.edu/about/diversity-equity-inclusion/tools-and-additional-resources/
glossary-dei-concepts.

UNL Physics 2018. Football physics with Dr. Tim Gay. Accessed 9 June 2018 at:
http://physics.unl.edu [T].

USA Science & Engineering Festival. 2018. Accessed 9 June 2018 at: https://
usasciencefestival.org.

USBJS (US Bureau of Justice Statistics). 2017. Total correctional population.
Accessed 18 December 2018 at: www.bjs.gov [T].

US Census Bureau. 2012. Nearly 1 in 5 people have a disability in the U.S. Census
Bureau reports. Accessed 26 December 2017 at: www.census.gov [T].

US Census Bureau. 2017a. U.S. and world population clock. Accessed 12 May 2017
at: www.census.gov [T].

US Census Bureau. 2017b. Quick facts: United States. Accessed 21 December 2017
at: www.census.gov [T].

US Department of Commerce. 2011. Economic and statistics administration
releases new report on STEM: good jobs now and for the future, 14 July 2011.
Accessed 8 May 2017 at: http://2010–2014.commerce.gov [T].

US Department of Education. 2017. Projected increases in STEM jobs: 2010–2020.
Accessed 27 March 2019 at: .www.ed.gov/sites/default/files/stem-
overview.pdf.

US Department of Education. 2018. Community of practice. Accessed 8 July 2018 at:
www.ed.gov [T].

Van Duzor, A. G. & M. Sabella. 2012. Science education internships for the
professional development of Noyce Scholars at Chicago State University:
affordances and constraints. Physics Education Research Conference. Accessed
20 April 2018 at: www.compadre.org [T].

Van Norden, R. 2013. Open access: the true cost of science publishing. *Nature
News*, 495:426–429.

Van Norden, R. 2014. On line collaboration: scientists and the social network.
Nature. Accessed 12 December 2017 at: www.nature.com [T].

van Weijen, D. 2012. The language of (future) scientific communication. Accessed
15 May 2017 at: www.researchtrends.com [T].

Vermeij, G. 1996. *Privileged Hands: A Scientific Life*. San Francisco, CA:
W. H. Freeman and Company.

Vicens, Q. & P. E. Bourne. 2007. Ten simple rules for a successful collaboration.
PLoS Computational Biology, 3:e44.

Vise, D. de. 2012. Colleges looking beyond the lecture. *Washington Post*. Accessed
2 September 2017 at: www.washingtonpost.com [T].

Visit Gainesville. 2018. Gainesville Solar Walk. Accessed 9 June 2018 at: www
.visitgainesville.com [T].

VSA (Visitor Studies Association). 2017. Accessed 9 December 2017 at: www
.visitorstudies.org.

Vogel, G. 2017. German researchers start 2017 without Elsevier journals. *Science*, 355:17.

Walhimer, M. 2011. How much do museum exhibitions cost? *Museum Planner*. Accessed 30 June 2018 at: https://museumplanner.org [T].

Walker, T. 2013. No more 'sit and get': rebooting teacher professional development. National Education Association. Accessed 9 April 2018 at: http://neatoday.org [T].

Walker, W. H. 2018. My second life as a teacher. *Science*, 359:246.

Waterfront Alliance. 2014. Dreaming of oysters. Accessed 23 June 2018 at: http://waterfrontalliance.org [T].

Wax, D. 2017. 10 tips for more effective PowerPoint presentations. Lifehack. Accessed 4 September 2017 at: www.lifehack.org/articles/featured/10-tips-for-more-effective-powerpoint-presentations.html.

WeDigBio. 2018. Accessed 28 June 2018 at: www.idigbio.org [T].

Weidler-Lewis, J., G. R. Lamb, & J. Polman. 2018. Using science infographics to jump-start creativity in the classroom. *Science Teacher*, 86(2):42–47

Weir, A. 2011. *The Martian*. New York: Broadway Books.

Welsh, J. 2012. Snake invader: images of Titanoboa in Grand Central. Accessed 31 December 2018 at: www.livescience.com.

Wenger, E., R. McDermott, & W. M. Snyder. 2002. *Cultivating Communities of Practice*. Boston, MA: Harvard Business Press.

Whang-Sayson, H., J. C. Daniel, & A. A. Russell. 2017. A serendipitous benefit of a teaching-exploration program at a large public university: creating a STEM workforce that supports teachers and public education. *Journal of College Science Teaching*, 47:24–30.

Whitemyer, D. 2018. 5 strategies for keeping your exhibit on budget. *Museum Next*. Accessed 9 June 2018 at: www.museumnext.com [T].

Whitmore, F. C. & R. H. Stewart. 1965. Miocene mammals and the Central American seaways. *Science*, 148:180–185.

Wiggins, A. & Y He. 2016. Community-based data validation practices in citizen science. In *Proceedings of the 19th Conference on Computer-Supported Cooperative Work & Social Computing* (pp. 1548–1559). San Francisco, CA, 27 February to 2 March 2016, DOI:10.1145/2818048.2820063.

Wikiquotes. 2018. Dwight D. Eisenhower. Accessed 18 November 2018 at: https://en.wikiquote.org [T].

Willcuts, M. P. H. 2009. *Scientist–Teacher Partnerships as Professional Development: An Action Research Study*. Washington, DC: US Department of Energy.

Wing, E. S. 1962. Succession of mammalian faunas on Trinidad, West Indies. Unpublished PhD dissertation, Department of Biology, University of Florida.

Wing, E. S. 2005. Zooarchaeology at FLMNH in the context of the growth and development of this discipline. *Bulletin of the Florida Museum of Natural History*, 44:11–16.

World Bank. 2017. Research and development expenditure (% of GDP). Accessed 9 May 2017 at: http://data.worldbank.org [T].

Wu, J. 2016. Innovation fact of the week: women represent 40% of STEM workforce in China, but only 24% in US. Accessed 28 November at: www .innovationfiles.org [T].

Yalowitz, S. S. & K. Bronnenkant. 2009. Timing and tracking: unlocking visitor behavior. *Visitor Studies*, 12:47–64, DOI:10.1080/10645570902769134.

Yang, D. 2013. Are we MOOC'd out?. *Huffington Post*. Accessed 14 April 2018 at: www.huffingtonpost.com [T].

Yoho, R. 2015. How science fairs shaped my career. *Science*, 349:1578.

Yoon, C. 1995. Scientist at work: Geerat Vermeij; getting the feel of a long-ago arms race. *New York Times*. Accessed 26 December 2017 at: www.nytimes.com [T].

You, J. 2014. The top 50 science stars of Twitter. *Science*. Accessed 16 September 2017 at: www.sciencemag.org/news [T].

YouTube. 2018. Bill Nye the Science Guy. Accessed 8 July2018 at: www.youtube.com [T].

Yuhas, A. 2015. How Republican presidential candidates are getting away with denying evolution. *Guardian*. Accessed 31 December 2018 at: www.theguardian.com.

Zaspel, J. 2017. Invertnet. Talk presented at the 2017 iDigBio Summit. Accessed 18 December 2017 at: www.idigbio.org [T].

Zhang. B. 2018. These are the busiest airports in the world. Accessed 31 December 2018 at: www.businessinsider.com.

Zooniverse. 2018. Accessed 24 June 2018 at: www.zooniverse.org.

Zubrin, R. 2015. How scientifically accurate is *The Martian*?. *Guardian*. Accessed 6 December 2017 at: www.theguardian.com [T].

Index